SCIENTIFIC EVIDENCE

SCIENTIFIC EVIDENCE

EVIDENCE

Philosophical Theories

& Applications

Edited by

PETER ACHINSTEIN

The Johns Hopkins University Press

Baltimore & London

The Johns Hopkins University Press
2715 North Charles Street
Baltimore, Maryland 21218-4363
www.press.jhu.edu

Library of Congress Cataloging-in-Publication Data

Scientific evidence : philosophical theories and applications /
edited by Peter Achinstein.
p.   cm.
Includes bibliographical references.
ISBN 0-8018-8118-8 (acid-free paper)
1. Science—Methodology—History.
2. Science—Philosophy—History. I. Achinstein, Peter.
Q174.8.S35 2005
501—dc22   2004023004

A catalog record for this book is available from the British Library.

For Linda

CONTENTS

Preface ix
Contributors xi

INTRODUCTION 1

 PART I

PHILOSOPHICAL THEORIES OF EVIDENCE

1/A LITTLE SURVEY OF INDUCTION 9
John D. Norton

2/FOUR MISTAKEN THESES ABOUT EVIDENCE,
AND HOW TO CORRECT THEM 35
Peter Achinstein

3/RESTORING AMBIGUITY TO ACHINSTEIN'S
ACCOUNT OF EVIDENCE 51
Steven Gimbel

4/THE FAST TRACK TO CONFIRMATION: ACHINSTEIN AND
PEIRCE ON EVIDENCE 69
Frederick M. Kronz and Amy L. McLaughlin

Contents

5 / POSITIVE RELEVANCE: A DEFENSE AND A CHALLENGE   85
Sherrilyn Roush and Peter Achinstein

Positive Relevance Defended / Sherrilyn Roush   85

A Challenge to Positive Relevance Theorists: Reply to Roush /
Peter Achinstein   91

6 / EVIDENCE AS PASSING SEVERE TESTS: HIGHLY PROBABLE
VERSUS HIGHLY PROBED HYPOTHESES   95
Deborah G. Mayo

7 / CONSILIENCE, CONFIRMATION, AND REALISM   129
Laura J. Snyder

PART II

SCIENTIFIC APPLICATIONS

8 / EVIDENCE FOR TRANSMUTATION IN SEVENTEENTH-
CENTURY ALCHEMY   151
Lawrence M. Principe

9 / AGENCY AND OBJECTIVITY IN THE SEARCH
FOR THE TOP QUARK   165
Kent W. Staley

10 / WILL GENOMICS DO MORE FOR METAPHYSICS
THAN LOCKE?   185
Alex Rosenberg

11 / IS DOMESTIC BREEDING EVIDENCE FOR (OR AGAINST)
DARWINIAN EVOLUTION?   207
Richard A. Richards

12 / EVIDENCE IN THE SCIENCES OF BEHAVIOR   237
Helen E. Longino

13 / INTROSPECTIVE EVIDENCE IN PSYCHOLOGY   259
Gary Hatfield

PREFACE

This volume is based on a conference on scientific evidence held in April 2003 at the Johns Hopkins University. All of the contributors to the volume were invited to attend the conference. The essays of Gimbel and Roush (and my reply to the latter) have been published elsewhere. The others are published here for the first time. Participants were asked to write an essay on an aspect of scientific evidence. Some chose to address general questions about the meaning of evidence. Others examined issues concerning evidence for particular scientific theories, with attention to the underlying concept of evidence employed. Because contributors were invited to write on problems of special interest to them, the volume includes discussions of only some concepts and views about evidence. It is hoped that the material presented here will stimulate others to think critically about the concepts of evidence discussed and to explore others not treated in the volume.

It is my pleasure to note that seven of the contributors are former students of mine who received their Ph.D. from the Johns Hopkins University at various times over the past 35 years. They are Steven Gimbel, Frederick Kronz, Helen Longino, Richard Richards, Alex Rosenberg, Laura Snyder, and Kent Staley. My own work on evidence has been advanced by reading and discussing their views, some of which contain sharp criticisms of my own. I count myself fortunate to have had students of this caliber.

The conference and this volume have been made possible by support from the Johns Hopkins Center for History and Philosophy of Science; by a bequest from my father, Asher Achinstein, to the Center; and by support from the Hopkins Philosophy Department and its chair, Michael Williams. Fred Kronz and Larry Principe helped to organize the conference. L. Suzanne Brown, to whom the volume is dedicated, was indispensable in making the conference possible. Finally, I owe a debt of thanks to Christopher Altermatt, Michael Hicks, Matthew Holtzman, Lisa Ievers, and Cherie McGill for valuable help with the proofs.

CONTRIBUTORS

Peter Achinstein, Professor of Philosophy at the Johns Hopkins University, is the author of numerous works in the philosophy of science on scientific method, explanation, probability, and evidence, including *Particles and Waves* (Lakatos Award, 1993) and *The Book of Evidence*.

Steven Gimbel, Associate Professor of Philosophy at Gettysburg College, has written papers in the philosophy of science, including "Unconventional Wisdom: Theory-Specificity in Reichenbach's Geometric Conventionalism" (*Studies in History and Philosophy of Science*) and "Peirce Snatching: Towards a More Pragmatic View of Evidence" (*Erkenntnis*).

Gary Hatfield is Adam Seybert Professor in Moral and Intellectual Philosophy at the University of Pennsylvania. He has written many articles on early modern philosophy and science, on the history and philosophy of psychology, and on theories of visual perception. He is the author of *The Natural and the Normative: Theories of Spatial Perception from Kant to Helmholtz* and *Descartes and the Meditations*.

Frederick M. Kronz, Professor of Philosophy at the University of Texas at Austin, has published papers on the philosophy of quantum mechanics and the theory of evidence. His current interests include philosophy of quantum field theory, non-standard theories of probability, and the philosophy of C. S. Peirce.

Helen E. Longino is leaving her post as Professor of Philosophy at the University of Minnesota to become Professor of Philosophy at Stanford University in 2005. The author of *Science as Social Knowledge* and *The Fate of Knowledge,* she is currently working on a philosophical analysis of scientific approaches to understanding human behavior, of which the chapter in this volume is a part.

Deborah G. Mayo, Professor of Philosophy at Virginia Polytechnic Institute and State University, is the author of *Error and the Growth of Experimental Knowledge* (Lakatos Award, 1998), and of many papers on topics of evidence and probability in the philosophy of science.

Amy L. McLaughlin is Assistant Professor of Philosophy at the Wilkes Honors College of Florida Atlantic University. In collaboration with Fred Kronz she has published a paper on Peirce in a collection entitled *Between Chance and Choice: Interdisciplinary Perspectives on Determinism*.

John D. Norton is Professor of History and Philosophy of Science at the University of Pittsburgh. He writes on induction and confirmation, and on the history and philosophy of physics, especially Einstein and his discovery of special and general relativity.

xii

Lawrence M. Principe is Professor in the Department of History of Science and Technology, the Department of Chemistry, and the Department of Philosophy at the Johns Hopkins University. He is the author of *The Aspiring Adept: Robert Boyle and His Alchemical Quest* and co-author of *Alchemy Tried in the Fire: Starkey, Boyle, and the Fate of Helmontian Chymistry*. Principe was the winner of the first Francis Bacon Award in the History of Science.

Richard A. Richards is Assistant Professor of Philosophy at the University of Alabama. His research interests are in the philosophy of biology. Published papers include "Darwin and the Inefficacy of Artificial Selection" (*Studies in the History and Philosophy of Science*), "Darwin, Domestic Breeding, and Artificial Selection" (*Endeavour*), and "Character Individuation in Phylogenetic Inference" (*Philosophy of Science*). Prior to his philosophical career, he performed with ballet companies in the United States and Europe.

Alex Rosenberg is R. Taylor Cole Professor of Philosophy at Duke University. He has written extensively in the fields of the philosophy of biology and the philosophy of science. Among his books are *The Structure of Biological Science*, *Economics: Mathematical Politics or Science of Diminishing Returns* (Lakatos Award, 1993), and *Philosophy of Science: A Contemporary Approach*.

Sherrilyn Roush, Assistant Professor of Philosophy at Rice University, writes on a range of topics in the philosophy of science. Two most recent articles are "Testability and the Unity of Science" (*Journal of Philosophy*) and "Testability and Candor" (*Synthese*). She has a forthcoming book entitled *Tracking Truth: Knowledge, Evidence, and Science*.

Laura J. Snyder is Associate Professor of Philosophy at St. John's University in New York City. She has published articles on scientific reasoning and evidence and is the author of a forthcoming book entitled *Reforming Philosophy: A Victorian Debate on Science and Society*.

Kent W. Staley, Assistant Professor of Philosophy at Saint Louis University, has recently published a book entitled *The Evidence for the Top Quark: Objectivity and Bias in Collaborative Experimentation*. He is also the author of papers on experimentation and evidence in *Philosophy of Science* and *Physics in Perspective*.

SCIENTIFIC EVIDENCE

INTRODUCTION

Scientists frequently disagree about whether, or to what extent, some set of data or observational results constitute evidence for a scientific hypothesis. Disagreements may be over empirical matters, such as whether the data or observational results are correct, or whether other relevant empirical information is being ignored. But conflicts also arise because scientists are employing incompatible concepts of evidence. Indeed, there are scientists who have explicitly formulated theories of evidence that conflict with ones formulated by other scientists. Let me mention three examples.

In the third edition (1726) of his great work *Principia,* Isaac Newton formulated four "Rules for the Study of Natural Philosophy," which he employed in deriving his law of gravity.[1] The rules are to be followed in empirical science to establish or confirm a scientific law. Two of the rules pertain to causal reasoning; the others, to inductive generalization. The causal rules require that one infer no more causes of a given phenomenon than are true and sufficient to explain the phenomenon, and that from effects of the same kind one should infer the same cause. The inductive rules say that if all observed bodies have a certain physical property, then one can infer that all bodies in the universe have it, and that propositions so inferred should be regarded as true until new observations show that the generalization must be altered. On Newton's view, then, one has supporting evidence for a law when one has observed phenomena from which one can infer a cause, which can then be generalized to all cases of a certain sort. This view of evidence we might call causal-inductive. Newton showed how each of his four rules of reasoning is invoked in deriving his law of gravity from observed phenomena pertaining to the orbits of the planets and their satellites.

This Newtonian conceptual view of evidence was spelled out in considerable detail in the nineteenth century by the philosopher John Stuart Mill, who carefully defined induction and gave methods for how to determine causes.[2] At the time Mill was writing, the wave theory of light had been widely accepted by scientists as much better supported by optical evidence than the rival particle theory. Using his version of the causal-inductive method, Mill rejected the idea that the wave theory had been highly confirmed, on the grounds that an unobservable ether supporting the waves was postulated by that theory, and no causal-inductive reasoning from optical observations for the existence of such an ether was forthcoming from wave theorists.

A quite different conception of scientific evidence was proposed in the mid-nineteenth century by William Whewell, a British scientist who wrote works on mineralogy and tides as well as on the history and philosophy of science.[3] Whewell explicitly rejected Newton's four rules of reasoning, as well as Mill's causal-inductivism, as being much too conservative and not permitting scientists to confirm hypotheses involving causes and other properties unlike any so far observed. According to Whewell, the scientist does not employ Newton's rules or Mill's methods, either in generating a hypothesis in the first place or in testing and confirming it once it has been tested. Scientists generate hypotheses by starting with observations and then by making a guess or conjecture—an act of mental creation that organizes the observations but is not an inference from them. Once the hypothesis is created, it is tested by determining whether three conditions are satisfied: (1) Does the hypothesis explain phenomena that have been observed and predict new ones? (2) Does it explain and predict phenomena of a kind different from the ones contemplated in its original formulation? (Whewell calls this a "consilience" condition.) (3) Over time, as more phenomena are observed, does the theoretical system of which the hypothesis is a part become simpler, more unified, more coherent? (This is Whewell's "coherence" condition.)

If the answer is yes to all three questions, then the hypothesis (and the containing theoretical system) is confirmed by all the observed phenomena. The observed phenomena constitute strong evidence that the hypothesis is true, even if the hypothesis is not derivable from the observed phenomena via a causal-inductive generalization. For example, Whewell believed that the wave theory of light (which both Newton and Mill rejected)—with its postulation of waves in an unobservable ether—was highly confirmed by observed optical phenomena, even though that hypothesis was not a causal-inductive generalization from those phenomena. In the twentieth century, the Whewellian holistic view came to be known as "inference to the best explanation." It was developed by a number of philosophers, including W. V. Quine,[4] Gilbert Harman,[5] and Peter Lipton.[6]

The third scientist I will mention who expressed a general conceptual view about evidence was the twentieth century physicist Richard Feynman.[7] It is a view that agrees with Whewell's on one point, namely, that a scientist generates a hypothesis in the first place by guessing it, not by inference from data. But then the disagreement with Whewell (as well as with Newton) becomes profound. Whewell and Newton believed that observed evidence is supposed to confirm a hypothesis, in the sense that it is supposed to show that the hypothesis is true or very probable. According to Feynman, evidence can only disconfirm a hypothesis, never confirm it. In seeking evidence for a hypothesis that has been guessed it is the job of the scientist to try to falsify the hypothesis, not verify it, because verification is impossible. The scientist does so by deriving observable consequences from the hypothesis using deductive reasoning. If any of the consequences is shown to be mistaken by observation and experiment, then the hypothesis must be rejected. Feynman writes:

> In general we look for a new law by the following process. First we guess it. Then we compute the consequences of the guess to see what would be implied if this law that we guessed is right. Then we compare the result of the computation to nature, with experiment or experience, compare it directly with observation, to see if it works. If it disagrees with experiment it is wrong. In that simple statement is the key to science. . . .

> You can see, of course, that with this method we can attempt to disprove any definite theory. If we have a definite theory, a real guess, from which we can conveniently compute consequences which can be compared with experiment, then in principle we can get rid of any theory. There is always the possibility of proving any definite theory wrong; but notice that we can never prove it right. Suppose that you invent a good guess, calculate the consequences, and discover every time that the consequences you have calculated agree with experiment. The theory is then right? No, it is simply not proved wrong.[8]

This view, which came to be known as *falsificationism,* was developed in considerable detail in a 1934 book in German by the philosopher Karl Popper.[9] Popper rejected inductivism, which sanctions inferring the truth or probability of a general hypothesis from observed instances; following

David Hume from the eighteenth century, he believed that inductive reasoning is rationally unjustified. What scientists do, and should do, Popper claimed, is try to falsify the strongest theory they can propose by deriving observable consequences from it. If these consequences are false, the theory must be rejected. If they are true, we can conclude that they provide evidence for the theory (which Popper called corroboration), but we cannot conclude that the theory is true, probable, or believable because the consequences are true. All we can say is that despite our best attempts to refute it the theory remains unfalsified.

Inductivism (Newton, Mill), explanationism (Whewell, Quine, Harman, Lipton), and falsificationism (Feynman, Popper) represent three of the most important views about evidence found in the writings of both scientists and philosophers of the present and the past. They yield conflicting concepts of scientific evidence, and thus conflicting views about whether, or to what extent, data constitute evidence for a scientific hypothesis, and whether, or to what extent, one can rationally believe the hypothesis on the basis of the data.

Let me mention two other views that were first developed in the twentieth century. One is Bayesianism (probabilism), a view espoused by some statisticians and philosophers of science.[10] The fundamental idea is that scientific evidence is to be understood completely by reference to mathematical probability, which is subject to precise rules. On the usual Bayesian theory, a body of observational results counts as evidence for a hypothesis if and only if it increases the probability of that hypothesis. This, although precise, yields a less-demanding concept of evidence than those of the inductivist and the explanationist. Data can increase the probability of a hypothesis even though the hypothesis is not a causal-inductive generalization of the data, and even though the hypothesis does not explain those data. In one respect, the Bayesian view is more stringent than that of the falsificationist because the former requires increase in probability and the latter does not. On the other hand, for the best evidence, the falsificationist requires that the scientist attempt and fail to falsify the strongest hypothesis; the Bayesian has no such requirement.

The views of evidence so far noted assume that there is some universal principle or set of principles of evidence, such as Newton's four rules, Whewell's consilience and coherence, or the Bayesian idea that evidence is governed by the rules of probability. These principles are treated as a priori necessary constraints, rather than as empirically justifiable ones. An opposing view, developed by Thomas Kuhn[11] and Paul Feyerabend[12] in the twentieth century, holds that there are no such universal rules of evidence. Kuhn asserted that rules of evidence are not universal but are different in each scientific "paradigm"—for example, different for Aristotle, Newton, and twentieth-century quantum physicists—and that even when criteria such as coherence and simplicity are invoked to support a hypothesis in the light of observations, these criteria are applied differently in different paradigms and even by scientists within the same paradigm. Feyerabend formulated an even more radical view expressed in his "principle of proliferation," according to which any rule of evidence, or of scientific method generally, should be violated by scientists in search of novel theories.

A very different position, which also rejects standard theories of evidence, is defended by John Norton in the present volume. It is the view that there are no a priori rules of evidence, only empirical ones. For example, what justifies some inductive generalization from the fact that all observed members of a population X have a property P to the hypothesis that all members of the population have it is not some a priori principle of induction, but instead empirical facts about population X and property P. Sometimes inductions of this sort are justified, sometimes not.

5

With this array of opposing views and concepts of evidence, it is no wonder that conceptual differences about evidence are at least partly responsible for differences over whether or to what extent some body of information counts as evidence for a hypothesis. The present volume is produced with the hope that theories concerning the concept of evidence and its application to scientific practice can be usefully explored by philosophers and historians of science.

NOTES

1. Isaac Newton, *The Principia: Mathematical Principles of Natural Philosophy,* trans. I. Bernard Cohen and Anne Whitman (Berkeley: University of California Press, 1999), 794–96.

2. John Stuart Mill, *A System of Logic* (London: Longmans, 1884).

3. William Whewell, *The Philosophy of the Inductive Sciences,* vol. 2, 1840 edition (London: Routledge, 1996).

4. W. V. Quine, *Word and Object* (Cambridge: MIT Press, 1960).

5. Gilbert Harman, "The Inference to the Best Explanation," *Philosophical Review* 74 (1965): 88–95.

6. Peter Lipton, *Inference to the Best Explanation* (London: Routledge, 1991, 2nd ed., 2004).

7. Richard Feynman, *The Character of Physical Law* (Cambridge: MIT Press, 1967).

8. Ibid., 156–57.

9. The book was first translated into English in 1959 with the title the *Logic of Scientific Discovery* (London: Hutchinson, 1959). It became very influential. Whether Feynman read this book is unclear. He makes no mention of it in his own book, from which the excerpt herein was taken.

10. For a very clear exposition of Bayesianism, see Colin Howson and Peter Urbach, *Scientific Reasoning,* 2nd ed. (Chicago: Open Court, 1993).

11. Thomas Kuhn, *The Structure of Scientific Revolutions,* 2nd ed. (Chicago: University of Chicago Press, 1970).

12. Paul Feyerabend, "Against Method: Outline of an Anarchistic Theory of Knowledge," in *Minnesota Studies in Philosophy of Science,* vol. 4, ed. Michael Radner and Stephen Winokur (Minneapolis: University of Minnesota Press, 1970), 17–130.

PART I

PHILOSOPHICAL THEORIES OF EVIDENCE

A Little Survey of Induction

John D. Norton

My purpose in this chapter is to survey some of the principal approaches to inductive inference in the philosophy of science literature. My first concern will be the general principles that underlie the many accounts of induction in this literature. When these accounts are considered in isolation, as is more commonly the case, it is easy to overlook the fact that virtually all accounts depend on one of very few basic principles and that the proliferation of accounts can be understood as efforts to ameliorate the weaknesses of those few principles. In the earlier sections, I lay out three inductive principles and the families of accounts of induction they engender. In later sections I review standard problems in the philosophical literature that have supported some pessimism about induction and suggest that their import has been greatly overrated. In the final sections I return to the proliferation of accounts of induction that frustrates efforts at a final codification and suggest that this proliferation appears troublesome only as long as we expect inductive inference to be subsumed under a single formal theory. If we adopt a material theory of induction in which individual inductions are licensed by particular facts that prevail only in local domains, then the proliferation is expected and not problematic.

Basic Notions

Inductive inference is the primary means through which evidence is shown to support the content of science. It arises whenever we note that evidence lends support to an hypothesis, in whatever degree, while not establishing it with deductive certainty. Examples abound. We note that some squirrels have bushy tails, and we infer that all do. We note that our cosmos is bathed in a 3K radiation bath, and this lends credence to the standard big bang cosmology that predicts it. Or we may infer to the limited efficacy of a drug if the health of members of a test group given the drug improves more than that of a control group denied the drug. This notion of induction is called *ampliative*, which means that the hypotheses supported are more than mere reformulations of the content of the evidence. In the simplest case, evidence that pertains to a small number of individuals is generalized to all. It is called ampliative because we amplify a small number to all. This modern tradition differs from an older tradition that can be traced back to Aristotle, in which induction meant generalization and the generalization need not necessarily be ampliative. In that tradition, taking a finite list (iron conducts electricity; gold conducts electricity . . .) and summarizing it (all metallic elements conduct electricity) is called a *perfect induction*, even though it is fully deductive.

In cases in which we find the evidence so compelling that we accept the hypothesis into our corpus of belief, we tend to use the terms *induction* or *inductive inference*. If the bearing of the evidence is weak and we merely want to register that it has lent some support to the hypothesis, then we more commonly use *confirmation*. When the evidence speaks against an hypothesis but does not disprove it, we count it as a case of *disconfirmation*. The introduction of the term *confirmation* into the inductive inference literature is relatively recent and is closely associated with the representation of inductive relations as probabilistic relations. If the degree of confirmation rises to a level at which we are willing to adopt the hypothesis without proviso, then we say that the hypothesis is detached from the evidence; if it is wanted, a formal theory of confirmation requires a *rule of detachment* to implement it. For detachment in a probabilistic account of induction, the rule would typically require the probability to rise above a nominated threshold.

THREE PRINCIPLES AND THE FAMILIES THEY ENGENDER

Any account of inductive inference specifies when an item of evidence inductively supports an hypothesis and, in some cases, provides a measure of the degree of support. There are so many such accounts, some of them with histories extending to antiquity, that it is impossible to discuss them all. The task of comprehending them is greatly simplified, however, once we recognize that virtually all accounts of induction are based on just three ideas. As a result, it is possible to group virtually all accounts of induction into three families. This system is summarized in Table 1.1. Each family is governed by a principle upon which every account in each family depends. I also list what I call the *archetype* of each family, which is the first use of the principle and a familiar account of induction in its own right. These archetypes suffer weaknesses, and the family of accounts grows through embellishments intended to ameliorate these weaknesses.

Table 1.1   Three Families of Accounts of Inductive Inference

Family	Inductive Generalization	Hypothetical Induction	Probabilistic Induction
Principle	An instance confirms the generalization.	The ability to entail the evidence is a mark of truth.	Degrees of belief are governed by a numerical calculus.
Archetype	Enumerative induction	Saving the phenomena in astronomy	Probabilistic analysis of games of chance
Weakness	Limited reach of evidence	Indiscriminate confirmation	Applicable to non-stochastic systems

11

Each of the three families is discussed in turn in the three sections to follow, and the entries in this table are explicated.[1] While most accounts of inductive inference fit into one of these three families, some span across two families. Achinstein's (2001) theory of evidence, for example, draws on ideas from both hypothetical induction and probabilistic induction, insofar as it invokes both explanatory power and probabilistic notions. Demonstrative induction, listed here under inductive generalization, can also be thought of as an extension of hypothetical induction.

INDUCTIVE GENERALIZATION

The Archetype: Enumerative Induction

The most ancient form of induction, the archetype of this family, is *enumerative induction*, or *induction by simple enumeration*. It licenses an inference from "some As are B" to "all As are B." Examples are readily found in Aristotle.[2] Traditionally, enumerative induction has been synonymous with induction, and it was a staple of older logic texts to proceed from deductive syllogistic logic to inductive logics based on the notion of enumerative induction. These include variant forms of enumerative induction such as *example* (this A is B; therefore that A is B) and *analogy* (a has P and Q; b has P; therefore b has Q). One variant form is quite subtle. Known as intuitive induction, it requires that induction must be accompanied by a "felt certainty on the part of the thinker" (Johnson, 1922: 192). Elusive as this notion is, it may have been tacitly part of the notion of enumerative induction as far back as Aristotle. It has been just as traditional to vilify enumerative induction. Francis Bacon (1620, First Book, §105) has the most celebrated jibe:

> The induction which proceeds by simple enumeration is puerile, leads to uncertain conclusions, and is exposed to danger from one contradictory instance, deciding generally from too small a number of facts, and those only the most obvious.

Part of that scornful tradition has been the display of counterexamples devised to make the scheme appear as foolish as possible. My own view is that the scorn is misplaced and that the counterexamples illustrate only that any induction always involves inductive risk. Since the conclusion is never guaranteed at the level of deductive certainty, failures are always possible. The actual inductive practice of science has always used enumerative induction, and this is not likely to change. For example, we believe all electrons have a charge of -1.6×10^{-19} Coulombs, simply because all electrons measured so far carry this charge.

Extensions

The principal weakness of enumerative induction is its very limited scope. It licenses inference just from some As are B to all, and that scheme is too impoverished for most applications in science. Most of the embellishments of the scheme seek to extend the inductive reach of the evidence. In doing this they extract the governing principle from the archetype of enumerative induction: An instance confirms the generalization.

There are two avenues for expansion. The first reflects the fact that enumerative induction is restricted by the limited expressive power of the syllogistic logic in which it is formulated.

The principal burden of Hempel's (1945) "satisfaction" criterion of confirmation is to extend this notion of confirmation from the context of syllogistic logic to that of the richer first-order predicate logic. Hempel's theory in all detail is quite complicated, but its core idea is simple. Take an hypothesis of universal scope, for example $(x)(Px \rightarrow Qx)$ (i.e., "For all x, if x has P then x has Q.") The development of the hypothesis for a class of individuals is just what the hypothesis would say if these individuals were all that there were. The development of $(x)(Px \rightarrow Qx)$ for the class $\{a,b\}$ would be $(Pa \rightarrow Qa) \& (Pb \rightarrow Qb)$. Hempel builds his theory around the idea that the development is the formal notion of instance. Thus his account is based on a formal rule that asserts that hypotheses are confirmed by their developments.

Elegant as Hempel's account proved to be, it was still restricted by the inability of instances to confirm hypotheses that used a different vocabulary. So the observation of bright spots in telescopes or on cathode ray tube screens could not confirm hypotheses about planets or electrons. The second avenue of expansion seeks to remedy this restriction. One of the most important approaches is a tradition of eliminative methods most fully developed in the work of Mill in his *System of Logic* (1872, Book III, Ch. 7). These methods are intended to aid us in finding causes. I may find, for example, that my skin burns whenever I give it long exposure to sunlight and that it does not burn when I do not. Mill's "Joint Method of Agreement and Disagreement" then licenses an inference to sunlight as the cause of the burn. There are two steps in the inference. The first is a straightforward inductive generalization: From the instances reported, we infer that my skin always burns just when given long exposure to sunlight. The second is the introduction of new vocabulary: We are licensed to infer to the sunlight as the cause. So the methods allow us to introduce a causal vocabulary not present in the original evidence statements.

Mill's methods extend the reach of evidence because we use theoretical results to interpret the evidence. In Mill's case it is the proposition that a cause is an invariable antecedent. Glymour's (1980) "bootstrap" account allows any theoretical hypothesis to be used in inferring the inductive import of evidence. For example, we observe certain lines in a spectrograph of light from the sun. We use known theory to interpret that as being light emitted by energized helium, and so we infer that our star (the sun) contains helium, which is an instance of the hypothesis that all stars contain helium. What is adventurous in Glymour's theory is that the hypotheses of the very theory under inductive investigation can be used in the interpretation of the evidence. Glymour added clauses intended to prevent the obvious danger of harmful circularity; however, the ensuing criticism of his theory focused heavily on that threat.

In Glymour's bootstrap, we use theoretical results to assist us in inferring from the evidence to an *instance* of the hypothesis to be confirmed. In the limiting case, we infer directly to the hypothesis and simply do away with the inductive step that takes us from the instance to hypothesis itself. In the resulting "demonstrative induction" or "eliminative induction" or "Newtonian deduction from the phenomena," we find that the auxiliary results we invoke are sufficiently strong to support a fully deductive inference from the evidence to the hypothesis. Although this scheme may have little interest for those developing theories of inductive inference, it has very great practical importance in actual science, since it offers one of the strongest ways to establish an hypothesis. It has been used heavily throughout the history of science. Newton used it repeatedly in his *Principia*; it has also been used often in the development of quantum theory (see Norton, 1993; 2000). It should be stressed that demonstrative induction can never do away with the need for other inductive inference schemes. Since it is fully deductive, it never really allows us to infer beyond what we already know. Its importance lies in alerting us to cases in which the inductive risk that is needed to move from evidence to hypothesis has already been taken elsewhere. If we have already accepted the inductive risk taken in believing the auxiliary results in some other part of our science, we find that we need take no further inductive risks in inferring from evidence to hypothesis.

14

HYPOTHETICAL INDUCTION

The Archetype: Saving the Phenomena

This family of accounts of induction is based on a quite different principle: The ability of an hypothesis to entail the evidence deductively is a mark of its truth. The origins of explicit consideration of this principle lie in astronomy. Plato is reputed to have asked his students to find what combinations of perfect circular motions would save the astronomical phenomena. (See, for example, Duhem, 1969.) Whether celestial motions that save the phenomena were thereby shown to be the true motions became a very serious issue when Copernicus showed, in the sixteenth century, that the astronomical motions could be saved by the motion of the earth. What resulted was a fierce debate, dragging into the seventeenth century, over whether these motions were nonetheless merely mathematical fictions or true motions (see, for example, Jardine, 1984).

What became clear then was that *merely* saving the phenomena was not enough. One could always concoct some odd hypothesis able to save the phenomena without thereby having a warrant to it. We can now present the problem in a compact and forceful way through the example of frivolous conjunction. Assume that some hypothesis H is able to entail deductively (i.e., save) the evidence E, usually with assistance from some auxiliary hypotheses. Then it is a simple matter of logic that a logically stronger hypothesis H′ = H&X has the same ability, even though the hypothesis X conjoined to H can be the most silly irrelevance you can imagine. If saving the phenomena is all that matters, then we are licensed to infer from E to H and just as much to H′, even though E now indirectly supports the silly X.

A principle had been extracted from the archetype of saving the phenomena in astronomy: The ability of an hypothesis to entail the evidence is a mark of its truth. But the unaugmented principle accorded the mark too indiscriminately, so a simple hypothetico-deductive account of induction—one that merely requires the hypothesis to save the evidence[3]—is not a viable account of induction. This one problem, directly or indirectly, drives the literature in this family. It seeks to embellish the simple account with some additional requirement that would tame its indiscriminateness.

Extensions

The most straightforward embellishment produces what I shall call *exclusionary accounts*. In them, we require that the hypothesis H entail the evidence E, and moreover, that there is some assurance that E would not have obtained had H been false. Thus competitors to H are excluded. In the simplest version we merely require that the evidence E, in conjunction with suitable auxiliaries, entails the hypothesis deductively. This is immediately recognizable as the *demonstrative induction* introduced earlier, although now the alternative term *eliminative induction* is more appropriate because we recognize the power of the inference to eliminate alternative hypotheses. Although this kind of deductive exclusion is less commonly possible, one can often show something a little weaker: that, were H false, then *most probably* E would not have obtained. This circumstance arises routinely in the context of controlled studies. Randomization of subjects over test and control group is designed to assure us that any systematic difference between test and control group (the evidence E) must be due to the treatment (the hypothesis H), because if H were false, the differences could arise only by vastly improbable coincidences. This model of traditional error statistical analysis drives such accounts of induction as Giere (1983) and the more thorough account of Mayo (1996).

15

Other exclusionary accounts draw on our quite vivid intuitions concerning quite vague counterfactual possibilities. Nearly a century ago, Perrin found that roughly a dozen independent experimental methods for determining Avogradro's number N all gave the same result. In what Salmon (1984, Ch. 8) calls a "common cause argument," Perrin argued that this is powerful evidence for the reality of atoms, for—and here is the counterfactual supposition—were atoms not real, it would be highly unlikely that all the experimental methods would yield the same value.[4] Related accounts of induction have been developed under the rubric of "common origin inferences" (Janssen, 2002) and Whewell's "consilience of induction."

A different approach attempts to tame the indiscriminateness of hypothetico-dedecutive confirmation by using the notion of simplicity. Of the many hypotheses that save the phenomena, we are licensed to infer to the simplest (see, for example, Foster and Martin, 1966, part III). This may seem a fanciful approach given the difficulty of finding principled ways to discern the most simple. In practice, however, it is much used. The most familiar usage comes in curve fitting. Although many curves may be adequate to the data points, allowing for experimental error, we routinely infer to the simplest. What makes a curve simpler is usually quite precisely specified; in the family of polynomials, the simpler are those of lower order. The preference for simplicity is so strong that standard algorithms in curve fitting will forgo a more complicated curve that fits better in favor of a simpler curve that does not fit the data as well.

In an account of inductive inference known as *abduction* or *inference to the best explanation* (Harman, 1965; Lipton, 1991), we tame the indiscriminateness of simple hypothetico-deductive inference by requiring the hypothesis not just to entail the hypothesis, but to explain it. We infer to the hypothesis that explains it best. The 3K cosmic background radiation is not just entailed by Big Bang cosmology; it is also quite elegantly explained by it. We would prefer it to another cosmology that can only recover the background radiation by artful contrivance that we do not find explanatory. Just as choosing the simplest hypothesis threatened to enmesh us in a tangled metaphysics of simplicity, this account brings us the need to explicate the notion of explanation. There is already quite an expansive literature on the nature of explanation; in the context of abduction, causal explanation—explaining by displaying causes—seems favored. In practical applications, identifying the simplest explanation can often be done intuitively and without controversy. One gets a sense of the dangers lurking if one considers a putatively successful experiment in telepathy. A parapsychologist would find the experimental result best explained by the truth of telepathy. A hard-boiled skeptic (like me) would find it best explained by some unnoticed error in the experimental protocols.

In what I call *reliabilist accounts*, merely knowing that an hypothesis saves the evidence is insufficient to warrant support for it from the evidence. In addition, we have to take into account how the hypothesis was produced. It must be produced in the right way; that is, it must be produced by a method known to be reliable. One of the best-known reliabilist accounts is incorporated in Lakatos' (1970) "methodology of scientific research programs." According to this methodology, theories arise in research programs and continued pursuit of the program is warranted by the fecundity of the program. A program scoring successful, novel predictions is "progressive" and worthy of pursuit, whereas one without them is languishing—"degenerating"—and might be abandoned. Decisions on theory evaluation must be made in the context of this history. Merely noting a static relationship between the theory and a body of evidence is not enough. This structure is already in place in Popper's (1959) celebrated account of scientific investigation pro-

16

ceeding through a cycle of conjecture and refutation. The newly conjectured hypothesis that survives serious attempts at falsification is "well corroborated," a status that can only be assigned in the light of the history of the hypothesis' generation. As part of his complete denial of inductive inference, Popper insisted that the notion of corroboration was quite different from confirmation. I have been unable to see a difference sufficient to warrant the distinct terminology.

Reliabilist approaches permeate other assessments of the import of evidence. We routinely accept the diagnostic judgments of an expert, even when we laypeople cannot replicate the expert's judgments from the same evidence. We do this because we believe that the expert arrived at the assessment by a reliable method, perhaps learned through years of experience. Reliabilism also underwrites our scorn for *ad hoc* hypotheses. These are hypotheses that are explicitly cooked up to fit the evidence perfectly. For example, in response to failed experiments to detect any motion of the earth in the light-carrying ether, we might hypothesize that the earth just happens to be momentarily at rest in it. We do not doubt that the hypothesis entails the evidence, but we doubt that the evidence gives warrant for belief in the hypothesis precisely because of the history of the generation of the hypothesis.[5]

17

PROBABILISTIC INDUCTION

The Archetype: Games of Chance

Accounts in this third family owe their origins to an advance in mathematics: the development of the theory of probability, starting in the seventeenth century, as a means of analyzing games of chance. It was recognized fairly quickly that these probabilities behaved like degrees of belief, so that the same calculus could be used to govern degrees of belief in inductive inference. The best way to represent the inductive import of evidence E on hypothesis H was to form the conditional probability $P(H|E)$, the probability of the hypothesis H given that E is true. One of the most useful results in the ensuing inductive logic was presented by Bayes:

$$P(H|E) = (P(E|H)/P(E)) \, P(H)$$

It allows us to assess how learning E affects our belief in H, for it relates $P(H)$, the prior probability of H, to $P(H|E)$ the posterior probability of H conditioned on E. These quantities represent our degree of belief in H before and after we learn E. To compute the change, Bayes' formula tells us, we need know only two other probabilities. $P(E)$ is the prior probability of

E, our prior belief in the evidence E obtaining whether we know H is true or not. The likelihood P(E|H) is often readily at hand. Loosely, it tells us how likely E would be if H were true; in case H entails E, the probability is one. For more details of this Bayesian approach to inductive logic, see Howson and Urbach (1993) and Earman (1992).

This probabilistic approach is at its strongest when we deal with stochastic systems, that is, those in which probabilities arise through the physical properties of the systems under consideration. These physical probabilities are sometimes called *chances*. In such cases we might well be excused for failing to distinguish our degree of belief from the chance of, say, the deal of a royal flush in a game of poker, or the decay of a radioactive atom sometime during its half-life. The weakness lies in the difficulty of knowing how to proceed when our beliefs pertain to systems not produced by stochastic processes. Why should simple ignorance be measured by degrees that conform to a calculus devised for games of chance? Why should I even expect all my many ignorances to admit the sort of well-behaved degrees that can always be represented by real numbers between 0 and 1?

18

Extensions

The archetype of the family is the probabilistic analysis of games of chance. Although it is not clear that exactly this archetype can be applied universally, a family of accounts has grown based on the notion that this archetype captured something important. It is based on the principle that belief comes in degrees, usually numerical, and is governed by a calculus modeled more or less closely on the probability calculus. The weakness addressed by the different accounts of the family is that degrees of belief are not the chances for which the calculus was originally devised.

There have been two broad responses to this weakness. The first is the majority view among those in philosophy of science who use probabilities to represent beliefs. They urge that, whereas chances may not be degrees of belief, the latter should nevertheless always be governed by the same calculus as chances. In support of this view, they have developed a series of impressive arguments based loosely on the notion that, if our beliefs are to conform to our preferences and choices, then those beliefs must conform to the probability calculus. The best-known of these arguments are the "Dutch book" arguments of de Finetti. In a circumstance in which we place bets on various outcomes, they urge that, were our degrees of belief not to conform to the probability calculus, we could be induced to accept a combinations of bets that would ensure a loss—a Dutch book[6] (see de Finetti, 1937; Savage, 1972).

This majority view has traditionally been accompanied by zealous efforts to give a precise interpretation of the notion of probability employed. Many candidates emerged and could be grouped into roughly three types: a physical interpretation based on relative frequencies of outcomes; a logical interpretation representing probabilities as degrees of entailment; and a subjective interpretation representing probabilities as conventionally chosen numbers constrained by the probability calculus (for a recent survey, see Gillies, 2000). Although an appeal for precision in meaning is important, in my view these demands became excessive and placed impossible burdens on the analysis that cannot typically be met by accounts of central terms in other theories. A demand that probabilistic belief be defined fully by behaviors in, for example, the context of betting is reminiscent of long-abandoned efforts to define all physical concepts in terms of measuring operations, or psychological states in terms of behaviors. Similarly, efforts to find a non-circular definition for probability in terms of relative frequencies seem to neglect the obvious worry that all terms cannot be given explicit, non-circular definitions without reintroducing circularity into the total system.[7] One can only guess the disaster that would have ensued had we made similarly stringent demands for precise definitions of terms such as the state vector of quantum theory, or energy in physics.

The second broad response to the weakness of the family is to accept that degrees of belief will not always conform to the probability calculus, and that weaker or even alternative calculi need to be developed. Take, for example, some proposition A and its negation −A, and imagine that we know so little about them that we just have no idea which is correct. On the basis of symmetry, one might assign a probability of $1/2$ to each. But that, roughly speaking, says that we expect A to obtain one in two times, and that seems to be much more than we really know. Assigning, say, 0.1 or even 0 to both seems more appropriate. But that is precluded by the condition in probability theory that the probabilities of A and −A sum to one. The simplest solution is to represent our belief state not by a single probability measure, but by a convex set of them.[8] More adventurously, one might start to construct non-additive theories, such as the Shafer-Dempster theory (Shafer, 1976). Once one starts on this path, many possibilities open—not all of them happy. Zadeh's "possibility theory," for example, computes the possibility of the conjunction (A and B) by simply taking the minimum of the possibilities of each conjunct. The loss of information is compensated by the great ease of the computational rule in comparison with the probability calculus (for a critique, see Cheeseman, 1986). My contribution to this literature is the "theory of random propositions." It is a strict weakening of the probability calculus with well defined semantics designed to show that alternative calculi can solve some of the notorious problems of the Bayesian system. The theory of random propositions, for example, is not prone to the problem of the priors; assigning a zero prior probability no longer forces all of its posterior probabilities to be zero.

PROPERTIES AND TENDENCIES

Once we identify these three families, it is possible to see some properties and tendencies peculiar to each. One of the most important is that inference in the family of inductive generalization tends to proceed from the evidence to the hypothesis: We start with "some As are B" and infer to "all As are B." In the family of hypothetical induction, the direction of inference is inverted. We first show that one can pass deductively from the hypothesis to the evidence, and then we affirm that the evidence supports the hypothesis. Therefore, assigning evidence to the depths and theory to the heights, I label inductive generalization as "bottom-up" and hypothetical induction as "top-down."

20

This difference inclines the two families to very different attitudes toward induction. In inductive generalization the distance between evidence and theory is small, so there is optimism about the power of evidence. The passage from evidence to theory is almost mechanical. We replace a "some" by an "all," or we methodically collect instances and apply Mill's methods to reveal the cause. This family is hospitable to logics of discovery, that is, to recipes designed to enable us to extract theoretical import from evidence. In hypothetical induction, on the other hand, the distance between evidence and theory is great. There is no obvious path suggested by the evidence to the hypothesis that it supports. We cannot pass from 3K cosmic background radiation to Big Bang cosmology by a simple rule. This invites a focus on a creative element in confirmation theory: We can confirm hypotheses only if we are independently creative enough to find the ones able to entail the evidence in the right way. Thus, this family is the more traditional home of pessimism over the reach of evidence and the underdetermination thesis to be discussed below. It also invites skepticism about the possibility of logics of discovery.

There is also an interesting difference between, on the one hand, inductive generalization and hypothetical induction and, on the other, probabilistic induction. In the former, there tends to be little sophistication in justifying the particular schemes. They are either taken to be self-evident, much as *modus ponens* in deductive logic is not usually justified, or they are made plausible by displaying case studies in which the scheme at issue is seen to give intuitively correct results. The justification of schemes is a great deal more elaborate and sophisticated in probabilistic induction, with serious attempts to support the chosen system that go beyond mere invocation of intuitive plausibility, such as the Dutch book arguments mentioned earlier.

The Problems of Induction

Stated

Science is an inductive enterprise, and its unmatched success must be counted as a triumph of inductive inference. Yet philosophers have traditionally found it hard to participate in the celebration. The reason is that inductive inference has been the customary target of skeptical philosophical critiques. Here I sketch three of the best known and indicate why I think none of them presents difficulties anywhere near as severe as is commonly assumed.

1. Hume's problem of induction, also known as *The problem of induction* (Salmon, 1967: 11). This problem, whose origin traces at least to Hume, identifies a difficulty in justifying any inductive inference. It asserts that any justification of a given inductive inference necessarily fails. If the justification is deductive in character, it violates the inductive character of induction. If it uses inductive inference, then it is either circular, employing the very same inference form in the justification, or it employs a different inductive inference form, which in turn requires justification, thereby triggering an infinite regress. So we cannot say that our next inductive generalization will likely succeed because our past inductive generalizations have succeeded, for that is an induction on our past successes that uses exactly the inductive scheme under investigation.

2. Grue (Goodman, 1983). In this problem, an application of instance confirmation purports to show that the observation of green emeralds can confirm that emeralds as yet unobserved are blue. The trick is to define the predicate *grue*, which applies to emeralds if they are observed to be green prior to some future time *t*, but are blue otherwise. That some emeralds were observed to be green is equally correctly described as the observation of grue emeralds. The observation confirms that all emeralds are green; in its redescribed form, it confirms that all emeralds are grue, that is, that emeralds yet unobserved are blue. One might be inclined to block the confirmation of "all emeralds are grue" by complaining that "grue" is a bogus, compound property not amenable to inductive confirmation. Goodman argues, however, that this complaint fails. He takes grue and an analogously defined "bleen" and uses them as his primitives properties. Green

and blue are defined in terms of these properties by formulae just like those used for grue and bleen, so they now appear equally bogus. There is a perfect symmetry in their mutual definitions.

3. The underdetermination thesis (Newton-Smith, 2001). In the case of cutting-edge science, we are used to several theories competing, with none of them decisively preferred, simply because sufficient evidence has not yet been amassed. The underdetermination thesis asserts that this competition is ineliminable, even for mature sciences. It asserts that any body of evidence, no matter how extensive, always fails to determine theory. The thesis is grounded largely in the remark that many distinct theories can save any given body of evidence, and in the possibility of displaying distinct theories that share the same observational consequences.

Answered

Let me briefly indicate why none of these problems is so severe. Although Hume's problem has generated an enormous literature, what has never been established is that inductive inference is the sort of thing that needs to be justified. If the very notion of a mode of justification is to be viable, we cannot demand that all modes must in turn be justified, on pain of circularity or infinite regress. Some modes must stand without further justification and inductive inference, as a fundamental mode of inquiry, seems as good a candidate as any. While I believe that settles the matter, some may find this response to be question-begging.[9] In the discussion that follows I suggest another, less expected, resolution.

In practice, gemologists have not been surprised by the continuing green color of emeralds. The reason is that they recognize, perhaps tacitly, that green is a natural kind property that supports induction, whereas grue is not.[10] What of the symmetry of grue/bleen and green/blue? Our judgment that green only is a natural kind term breaks the symmetry. We might restore it by extending the "grue-ification" of all predicates until we have a grue-ified total science. We now have two sciences, with green a natural kind term in one and grue a natural kind term in the other. However, I have urged elsewhere that, with this extension, we now have two formally equivalent sciences and cannot rule out that they are merely variant descriptions of the very same facts. Then the differences between green and grue would become an illusion of different formulations of the same facts (see Norton, forthcoming).

If the underdetermination thesis is intended to pertain to the reach of evidence by inductive inference, its continued popularity remains a puzzle, for it can only be viable as long as one ignores the literature in induction and confirmation. It seems to be based essentially on the notion that many theories can save the same phenomena, and thus that they are equally confirmed by it. But as we saw in some detail in the discussion of hypothetical induction, that notion amounts to the simple hypothetico-deductive account of confirmation that has essentially never been admitted as a viable account—precisely because of its indiscriminateness. The underdetermination thesis presumes that this indiscriminateness is the assured end of all analysis and neglects the centuries of elaboration of accounts of hypothetical induction that followed, not to mention accounts in the other two families. I have also argued that the possibility of observationally equivalent theories proves to be a self-defeating justification for the thesis. If the equivalence is sufficiently straightforward to be established in philosophical discourse, then we can no longer preclude the possibility that the two theories are merely notational variants of the same theory. In that case, the underdetermination ceases to pertain to factual content (see Norton, manuscript).

THE NATURE OF INDUCTION
The Problem of Proliferation

Although I have reviewed some of the major problems of induction, there is another neglected problem implicit in the survey. We think of induction as a sort of logic, but it is quite unlike deductive logic in at least one important aspect. Deductive logic proceeds from a stable base of universal schemas. There is little controversy over *modus ponens*: If A then B; A; therefore B. Any grammatically correct substitution for A and for B yields a valid deduction. In contrast, the literature on induction and confirmation has a proliferation of accounts, and there seems to be no hope of convergence. Even our best accounts seem to be in need of elaboration. We are happy to infer to the best explanation, but delimiting precisely which inferences that licenses must await further clarification of the nature of explanation and the means for determining what explains best. The Bayesian system works well when our beliefs pertain to stochastic processes, but we are less sure we have the right calculus when the objects of belief stray far from such processes. Applying standard schemes can reveal gaps, even in simple cases. We measure the melting point of a few samples of the element bismuth as 271° C. On the strength of these measurements, by enumerative induction, we feel quite secure in inferring that all samples of bismuth have this melting point. But had our observation been of the melting point of a few samples of wax, or of the colors of a few birds of a new species, we would be very reluctant to infer to the generalization. In short, our accounts of induction are proliferating and, the more closely we look at each, the more we are inclined to alter or add further conditions. Induction seems to have no firm foundation, or if it has, we seem not to have found it.

Has this problem already been solved by one of the accounts outlined above? If there is one account of induction that does aspire to be the universal and final account, it is the Bayesian account. Although its scope is great and its ability to replicate almost any imaginable inductive stratagem is impressive, let me briefly indicate why it cannot yet claim this universal status. There are two problems. First, probabilities are not well adapted to all situations. For example, they are additive measures, and we know from measure theory that some sets are unmeasurable. We can contrive circumstances in which we seek a degree of belief in an unmeasurable set (see Norton, 2003, Section 3.4).

Second, while it is true that Bayesianism has an impressive record of replication of existing inductive stratagems, this record comes at a cost. We end up positing a fairly large amount of additional, hidden structure somehow associated with our cognition: Our beliefs must behave like real numbers or sets of them; we must harbor large numbers of likelihoods and prior probabilities; and we must somehow tacitly be combining them in just the way the probability calculus demands, even though carrying through the calculation explicitly with pencil and paper might well be taxing, even for an accomplished algebraist. We might not demand that these extra structures be present tacitly in our actual thought processes. We might merely expect a formal analysis using these structures to vindicate our inductive stratagems. There would still be a difficulty, insofar as these extra structures can extend too far beyond the structures of our actual stratagems. This is reminiscent of the situation in geometry a century ago when non-Euclidean geometries began to break into physics. Whatever the geometrical facts, one could always restore Euclidean geometry by adding further hidden geometrical structure. A three-dimensional spherical geometry, such as Einstein introduced with his 1917 cosmology, could be constructed by taking a hyperspherical surface in a four dimensional Euclidean space. That a Euclidean geometry could accommodate this new geometry was not decisive; it could only do it at the cost of adding additional structure (an extra spatial dimension) that was deemed physically superfluous. Analogously, we need to assess whether our gains from the more adventurous applications of Bayesianism are outweighed by the cost of the additional structures supposed. A Bayesian analysis serves two masters. The structures it posits must conform both to our good inductive stratagems and to the probability calculus. Those dual demands are often answered by the positing of quite rich structures. We saw earlier that belief states of total ignorance were represented as convex sets of infinitely many probability measures, in order not to violate the additivity of the probability calculus. If we drop one of these two masters—that the demand that the structures posited must in the end conform to the probability calculus—then we are more likely to be able to develop an account that posits sparser structures, closer to those present in our actual inductive practices.

A Material Theory of Induction

I have argued in Norton (2003) that the problem of proliferation admits a simple solution. The problem arises because we have been misled by the model of deductive inference, and we have sought to build our accounts of induction in its image. That is, we have sought formal theories of induction based on universal inference schemas. These are templates that can be applied universally. We generate a valid inductive inference merely by filling the slots of a schema with terms drawn from the case at hand. The key elements are that the schemas are universal—they can be applied anywhere—and that they supply the ultimate warrant for any inductive inference.

If there is such a formal theory that fully captures our inductions, we have not yet found it. The reason, I propose, is that our inductions are not susceptible to full systematization by a formal theory. Rather, our inductions conform to what I call a "material theory of induction." In such a theory, the warrant for an induction is not ultimately supplied by a universal template; the warrant is supplied by a matter of fact.[11] We are licensed to infer from the melting point of a few samples of bismuth to all samples because of the fact that elements are generally uniform in these physical properties. That fact licenses the inference, but without making the inference a deduction. The inductive character is preserved by the qualification "generally." Contrast this with the case of waxes. We are not licensed to infer from the melting point of a few samples of wax to all samples because there is no corresponding, licensing fact. In general, waxes are different mixtures of hydrocarbons with no presumption of uniformity of physical properties. The general claim is that, in all cases, inductions are licensed by facts. I have given facts that bear this function the name *material postulate*.

The idea that the license for induction may derive from facts about the world is certainly not new. Perhaps its best known form is Mill's (1872, Book III, Ch. III) postulation of the "axiom of the uniformity of the course of nature." The difficulty with seeking any one such universal fact to license all inductions is that the fact must be made very vague if it is to pretend to have universal scope, so much so that it is no longer usable. As a result I urge that inductions be always licensed by material postulates that prevail only in local domains—specific facts about elements license inductions in chemistry, specific facts about quantum processes license inductions about radioactive decay, and so on. As a result, I urge that "all induction is local."

Identifying the Material Postulates

The examples of the melting points of bismuth and wax illustrate the principal assertions of the material theory of induction. What grounds do we have for believing that the theory holds for all inductive inference forms and not just the examples presented? Those grounds lie in a review of a representative sample of inductive inference forms to be presented here. We will see that each form depends upon one or another factual, material postulate. The ease with which they are identifiable and the way we will see them arising will make it quite credible that material postulates underlie all viable inductive inference forms.

A survey of earlier sections makes it easy for us to sample inductive inference forms from across the literature and affirm that they are all grounded in some sort of local material postulate. (For further discussion, see also Norton, 2003, §3.) The family of forms encompassed by inductive generalization is the easiest to analyze. All of the ampliative inductive inference forms in that family depend upon the same inductive move: An instance confirms the generalization. They differ only in the details of how the generalization is specified, with Hempel's account being the most elaborated. All agree that an individual that has property P confirms that all individuals have property P, where no restriction is placed on the selection of property P. As familiar problems such as grue have shown, that is far too permissive a condition. Further constraints must be placed on which properties P support generalization. These are routinely factual constraints, such as we saw in the case of bismuth and wax, and these added constraints also typically supply us information of great importance on the strength of the induction.

The basic principle of the family of forms embraced by hypothetical induction is that an hypothesis is confirmed by its deductive consequences. So the hypothesis A & B is confirmed by its consequence A. If the principle is to be non-trivial, that A confirms A & B must amount to more than the trivial A confirms A; it must mean that A confirms B as well. That is admissible only if there is some sort of relevance relation between A and B—for example, that truth of A makes B's truth more likely.[12] But there is no general theory of what it is for A to be relevant to B; it is introduced as a particular fact essential to the viability of the induction. This pattern continues as we review the various augmented forms that compose the family. In the exclusionary accounts, we have an added assurance that the evidence would probably not have obtained had the hypothesis been false. That assurance is supplied by factual conditions. For example, they are in the factual conditions describing a controlled experiment that assure us that the only systematic difference between test and control group is the application of the treatment. Or they reside in the speculative but widely held belief that, were matter not atomic, Perrin would not have recovered the agreements found in his multiple measurements of Avogadro's number. In another approach, we augment the simple hypothetico-deductive schema by the requirement that the hypotheses in question be simple. There is no universal, formal account of what it is to be simple. Judgments of simplicity, in practice, turn out to reflect our beliefs regarding the basic ontology of the domain in question. In curve fitting, the natural default is to judge a linear relation to be simpler than a polynomial. That reflects factual presumptions, and these presumptions are made apparent when our simplicity rankings are reversed merely by a change of domain. If we are fitting trajectories to newly observed celestial objects, we will be inclined to fit a conic section ahead of any other curve, even a straight line. Factually, the conic sections are the simplest curves in that new domain; according to Newton's theory of gravity, they are the trajectories of bodies in free fall.

27

This pattern persists in the remaining inductive inference forms in the family of hypothetical induction. In inference to the best explanation, we augment the hypothetico-deductive schema by the requirement that the hypothesis best explain the evidence. Judgments of what explains and what best explains are governed by factual matters. For example, take the imaginary experiment mentioned earlier that purports to vindicate telepathy. Judgments of what explains best are essentially determined by the facts we presume to prevail. The parapsychologist, inclined to believe in paranormal processes, explains the outcome of the experiment by the supposition of telepathic processes; the skeptic, disinclined to believe in them, explains it by an unnoticed procedural error. Finally, reliabilist accounts of induction add the requirement that the hypothesis that saves the evidence must be generated by a reliable method. The viability of reliabilist accounts depends essentially on an important factual assumption: that the world is such that methods used do actually lead to the truth. If we do not think that fact obtains, then no inductive gain arises from following the method of hypothesis generation that is touted as reliable. For this reason, most of us do not heed predictions of oracles and entrail readers, no matter how perfectly they follow their standard methods.

We saw earlier that probabilistic accounts of induction work best when our beliefs pertain to stochastic systems, for then we can adopt the physical chances as degrees of belief that automatically conform to the probability calculus. The facts that license the resulting inductions are just the facts that govern the stochastic processes. For example, the half-life of Radium 221 is

30 seconds, so the chance that a given atom will decay in 30 seconds is 0.5. That fact licenses the corresponding probabilistic degree of belief and, using the calculus to which probabilities conform, we infer that the atom will almost certainly decay over the next 300 seconds. The probability of decay is $1 - (0.5)^{10} = 0.999$. Matters are far less clear when our beliefs pertain to circumstances for which physical chances are unavailable. Then why should our beliefs be captured by numerical degrees and, if they can be, why should those numerical beliefs conform to the probability calculus? Sources mentioned in the section on probabilistic induction seek to argue that our beliefs in this extended domain should still be probabilities. These arguments proceed from factual premises, so it is a simple matter to convert each of these arguments to serve the material theory of induction. These factual premises supply the material postulate for a domain in which a probabilistic account of induction is appropriate. For example, the Dutch book arguments tell us that probabilistic induction is appropriate if we find ourselves in a situation in which we are prepared to accept bets according to the presumptions laid out. Where such arguments are available, we can identify facts that constitute the material postulate; where such arguments are not available, we have no good reason to expect probabilistic induction to be applicable.

More Generally

Seeing this number of examples gives a sense of inevitability for the possibility of identifying a material postulate for any inductive inference form. The identification can almost be reduced to a general recipe. For each inductive inference form, one merely needs to conceive a factual circumstance in which the form would fail. The presumption that this factual circumstance does not obtain is then declared to be part of the material postulate that licenses the induction.

We can strengthen our expectation that a material postulate can always be recovered by considering how they were recovered in the cases above. In each, the postulate remedies an incompleteness in the specification of the inductive inference form. There are two sorts of incompleteness. Some are incomplete in so far as they simply need factual presumptions for their applicability; there is no expectation that we will supplant these factual presumptions with a general theory. For example, the exclusionary accounts always require some facts to sustain the requirement that the evidence would be unlikely to obtain if the hypothesis were false. Or reliabilist accounts presume that the relevant methods are properly adapted to our world, which is a matter of fact about the methods and the world. Probabilistic accounts require that we be in a domain in which beliefs can properly be required to conform to the probability calculus. In these cases, straightforwardly, the inductive inference forms can be employed only when the pertinent facts obtain.

In the other sort of incompleteness, the full specification of the inductive inference form calls up general notions from other parts of the philosophy of science. Might these full specifications be universal in scope and thus provide counterexamples to the material theory of induction's claim that all induction is local? The envisaged counterexamples fail. Forming them requires general accounts of these extra notions drawn from other parts of the philosophy of science, but these general accounts have, in each case, proven elusive. Instead of seeking them, we cite facts to complete demonstration of the applicability of the inductive inference form to the case at hand. For example, in the case of inductive generalization, we need some way to select which properties support generalization, and to what degree. That might come from the notion of natural kind or from notions of physical laws in which those properties may appear as fundamental terms. In the absence of a compact account of natural kinds or physical laws, we cite facts to license the induction: Physical properties of elements support inductive generalization; grue-ified properties do not. For simple hypothetico-deductive confirmation, we require a general notion of relevance applicable to parts of a theory. In the absence of a general account of the notion of relevance, we may merely cite the fact that the parts of a theory are relevantly related. When the scheme is augmented by the requirement that the hypotheses to be confirmed must be simple, we do not resort to the ever-elusive, general theory of what it is to be simple. We simply declare what is simpler than what and, as we saw above, those declarations are governed by the facts prevailing in the case at hand.

Finally, in the case of inference to the best explanation, we lack an appropriate account of explanation that could be used to give a full specification of the inductive inference form. If explaining is merely subsumption under a covering law, then the form reverts to simple hypothetico-deductive confirmation. If explanation is the displaying of causes, then we face the problem that there is no generally agreed upon account of the nature of causation that would allow us to specify just what counts as a cause. In practice, however, we do not need a general account of explanation. In each case, we believe certain facts prevail and provide guides to the sorts of results that may govern the process at hand. We typically pick from this repertoire the hypotheses that we believe explain the evidence, and which do it best. In short, in all of these cases we may speak of general notions such as law, natural kind, simplicity, and explanation, but in practice we ground the induction not in the general theory of these notions, but in particular facts.

For these reasons, the material theory of induction urges that all inductive inference be ultimately licensed by facts prevailing only in local domains. Indeed, the theory presents a challenge to critics: Can we find a universally valid inductive inference schema? The material theory predicts that this is not possible.

The Problems It Solves

This locality of induction resolves the problem of proliferation. Facts in each domain license inductions unique to that domain. There are many cases in which inductions licensed in different domains appear similar. The inference from the melting point of some samples of bismuth to all samples is formally like the inference from the blackness of some ravens to all. We codify this with a schema: enumerative induction. But the similarity is superficial, because the two inductions are ultimately licensed by very different facts. The first induction is quite strong and the second somewhat weaker, since color uniformities in birds are far less reliably obtained than the uniformities of physical properties of elements. This difference of strength is a puzzle for the scheme of enumerative induction. It is expected according to a material theory, since the inductions are from different domains where they are licensed by different facts. As we restrict the domain further, the particular facts available to license induction become more specific and varied. As a result, we expect this sort of closer scrutiny to produce a greater variety of formally different sorts of inductions, with no presumption of a convergence to a few identifiable forms. This divergent behavior was identified as the problem of proliferation of formal theories. From the perspective of a material theory, the proliferation is an unproblematic artifact of the nature of induction.[13]

Finally, a material theory of induction casts a very different light on Hume's problem of induction. That problem, as we saw earlier, arises from our ability to generate difficulties rapidly in seeking a justification of an induction. Our justification is circular, insofar as we use the same inductive inference form for justification, or it triggers an infinite regress, insofar as we must always invoke another inductive inference form to justify the last one. The regress is problematic even in its early stages. It is fanciful to carry out a meta-induction on inductions, and still more fanciful to try to carry out meta-meta-inductions on meta-inductions on inductions—and that has taken us only two steps into the regress. A material theory of induction eludes the problem simply because it is impossible to set up the difficulty, if one accepts the characterization of induction supplied by the material theory. An induction is licensed by its material postulate, but since the material postulate is itself another fact within our science, there is little mystery in what licenses it. It is licensed by other inductions based on other material postulates, and those in turn are licensed by further inductions based on further material postulates. This is a regress, but it is not the philosopher's fanciful fiction of an ascent through meta-inductions, meta-meta-inductions, and so on. Rather, it is the real and mundane business of ordinary science that traces our justification for this fact in that fact, and for that fact in some further fact, and so on. One might wonder how the resulting chains might end. As an empiricist I expect them to end in experiences; if there is a problem to be found, one might expect it to be there. But a lot more needs to be said to establish a problem, and what is said in setting up the traditional version of Hume's problem does not say it. For more discussion, see Norton (2003, §6).[14]

ACKNOWLEDGMENT
I am grateful to Peter Achinstein for helpful discussion and comments on an earlier draft of this chapter and to Phil Catton for helpful discussion.

NOTES
1. It is hopeless to try to include every view on induction in a survey of this size. I apologize to those who find their views slighted or omitted.

2. For example, *Prior Analytics* Book II.23 68b15–20 (McKeon, 1941: 102); *Topics* Book I.12 105a10–19 (ibid., 198).

3. Here and henceforth I tacitly assume that the hypothesis entails the evidence, usually with the aid of auxiliary hypotheses.

4. This same argument form, specifically involving multiple experiments to measure some fundamental constant, has arisen in other areas, so I have given it the narrower label of the method of overdetermination of constants (Norton, 2000, §3).

5. Kelly (1996) has developed a general framework for reliabilist accounts using the theoretical apparatus of formal learning theory. The framework extends well beyond the simple notion here of reliabilism as an augmentation of hypothetico-deductivism.

6. The weakness of all these arguments is that one cannot recover something as highly structured as the probability calculus from an argument without including assumptions of comparable logical strength in its premises. These assumptions are introduced through presumptions about the context of our decisions, the structures of our preferences, and the rules for translating beliefs into actions. A skeptic recognizes that these assumptions are essentially generated by working backward from a foregone conclusion—that beliefs must in the end conform to the probability calculus—to a set of assumptions carefully contrived to encode just that calculus. The skeptic wonders why the entire exercise could not be abbreviated simply by positing at the start that beliefs are probabilities.

7. If an outcome has probability $1/2$, then in the limit of infinitely many repetitions the frequency of success will not invariably approach $1/2$; rather, it will approach $1/2$ *with probability* 1—so a definition based on this fact becomes circular.

8. For example, we might take $P_1(A) = 0.1$ $P_1(-A) = 0.9$ and $P_2(A) = 0.9$ $P_2(-A) = 0.1$ and choose as our set all probability measures that can be formed as suitably weighted averages of them: $P_\alpha = \alpha P_1 + (1 - \alpha)P_2$, where $0 \le \alpha \le 1$.

9. This is one of several standard answers to the problem. For an entrance into this quite enormous literature, see Earman and Salmon, 1992, §§2.5–2.6.

10. Again, this is one standard response drawn from an enormous literature (see Stalker, 1994).

11. This is not such a strange notion, even in deductive inference. An inference from Socrates' humanity to his mortality is warranted by the fact that all men are mortal. What is different in this example is that (unlike a material theory of induction) the warrant for the inference is in turn supplied by a schema, *modus ponens.*

12. This is a variation of the problem of frivolous conjunction, which is blocked by the factual assertion of relevance.

13. It is now apparent also why Bayesianism can have the broadest scope. The system is quite weak and is able to license little until a large number of parameters are set. These are the prior probabilities and the likelihoods. The settings of these parameters can be used to encode a lot of factual information about particular domains. The material theory predicts that inductive inferences are particularized to the domains in which they arise; the settings of these parameters enable the Bayesian system to adapt itself to each domain with great flexibility.

32

14. After this chapter was written, I was very grateful to Paul Griffiths and also to Samir Okasha for pointing out to me that the basic idea of a material theory of induction had already been outlined in Sober (1991, Ch. 2), "The Philosophical Problem of Simplicity." Sober urges that an hypothesis is only ever confirmed by evidence in relation to a background theory or background assumptions, so that there is no general principle of induction. His purpose was to elucidate how appeals to simplicity are used to sustain induction. This analysis showed that these appeals are indirect appeals to relevant empirical assumptions, as I also concluded in Norton, 2003, Section 3.3.2.

REFERENCES

Achinstein, Peter (2001) *The Book of Evidence.* New York: Oxford University Press.

Bacon, Francis (1620) *Novum Organum in Advancement of Learning: Novum Organum: New Atlantis.* Great Books of the Western World, vol. 30. Chicago: University of Chicago Press, 1952.

Cheeseman, Peter (1986) "Probability versus Fuzzy Reasoning," pp. 85–102 in L. N. Kanal and J. F. Lemmer, eds., *Uncertainty in Artificial Intelligence.* Amsterdam: North Holland.

De Finetti, Bruno (1937) "Foresight: Its Logical Laws, Its Subjective Sources." *Annales de l'Institut Henri Poincaré,* 7; reprinted in *Studies in Subjective Probability,* pp. 95–157, eds., H. E. Kyburg and H. E. Smokler, New York: Wiley.

Duhem, Pierre (1969) *To Save the Phenomena, an Essay on the Idea of Physical Theory from Plato to Galileo.* Trans. E. Doland and C. Maschler. Chicago: University of Chicago Press.

Earman, John (1992) *Bayes or Bust: A Critical Examination of Bayesian Confirmation Theory.* Cambridge, MA: Bradford.

Earman, John and Salmon, Wesley C. (1992) "The Confirmation of Scientific Theories," Ch. 2 in M. H. Salmon et al., *Introduction to the Philosophy of Science.* Indianapolis: Hackett, 1999 (reprint).

Foster, Margueritte H. and Martin, Michael L. (1966) *Probability, Confirmation, and Simplicity: Readings in the Philosophy of Inductive Logic.* New York: The Odyssey Press.

Giere, Ronald R. (1983) "Testing Theoretical Hypotheses," pp. 269–98 in J. Earman, ed., *Testing Scientific Theories: Minnesota Studies in the Philosophy of Science,* Vol. 10. Minneapolis: University of Minnesota Press.

Gillies, Donald (2000) *Philosophical Theories of Probability.* London: Routledge.

Glymour, Clark (1980) *Theory and Evidence.* Princeton: Princeton University Press.

Goodman, Nelson (1983) *Fact, Fiction and Forecast*, 4th ed. Cambridge, MA: Harvard University Press.

Harman, Gilbert (1965) "Inference to the Best Explanation." *Philosophical Review* 74, 88–95.

Hempel, Carl G. (1945) "Studies in the Logic of Confirmation." *Mind* 54, 1–26, 97–121; reprinted with changes, comments, and postscript (1964) in Carl G. Hempel, *Aspects of Scientific Explanation*. New York: Free Press, 1965.

Howson, Colin and Urbach, Peter (1993) *Scientific Reasoning: The Bayesian Approach*, 2nd ed. La Salle, IL: Open Court.

Janssen, Michel (2002) "COI Stories: Explanation and Evidence in the History of Science." *Perspectives on Science* 10, 457–522.

Jardine, Nicholas (1984) *The Birth of History and Philosophy of Science: Kepler's A Defence of Tycho against Ursus, with Essays on Its Provenance and Significance.* Cambridge: Cambridge University Press.

Johnson, W. E. *Logic Part II: Demonstrative Inference: Inductive and Deductive.* Cambridge: Cambridge University Press (reprinted, New York: Dover, 1964).

Kelly, Kevin T. (1996) *The Logic of Reliable Inquiry.* New York: Oxford University Press.

Lakatos, Imre (1970) "Falsification and the Methodology of Scientific Research Programmes," pp. 91–196 in I. Lakatos and A. Musgrave, eds., *Criticism and the Growth of Knowledge.* Cambridge: Cambridge University Press.

Lipton, Peter (1991) *Inference to the Best Explanation.* London: Routledge.

Mayo, Deborah (1996) *Error and the Growth of Experimental Knowledge.* Chicago: University of Chicago Press.

McKeon, Richard (1941) *The Basic Works of Aristotle.* New York: Random House.

Mill, John Stuart (1872) *A System of Logic: Ratiocinative and Inductive: Being a Connected View of the Principles of Evidence and the Methods of Scientific Investigation,* 8th ed. London: Longman, Green, and Co., 1916.

Newton-Smith, William H. (2001) "Underdetermination of Theory by Data," pp. 532–36 in W. H. Newton-Smith, ed., *A Companion to the Philosophy of Science.* Blackwell.

Norton, John D. (1993) "The Determination of Theory by Evidence: The Case for Quantum Discontinuity 1900–1915." *Synthese* 97, 1–31.

Norton, John D. (1994) "Theory of Random Propositions," *Erkenntnis* 41, 325–52.

Norton, John D. (2000) "How We Know About Electrons," in Robert Nola and Howard Sankey, eds., *After Popper, Kuhn and Feyerabend.* Kluwer, pp. 67–97.

Norton, John D. (2003) "A Material Theory of Induction." *Philosophy of Science* 70, 647–70.

Norton, John D. (manuscript) "Must Evidence Underdetermine Theory?" First Notre Dame-Bielefeld Interdisciplinary Conference on Science and Values, Zentrum für Interdisziplinäre Forschung, Universität Bielefeld, July 9–12, 2003. philsci-archive.pitt.edu

Norton, John D. (forthcoming) "How the Formal Equivalence of Grue and Green Defeats What Is New in the New Riddle of Induction." *Synthese.*

Popper, Karl R. (1959) *Logic of Scientific Discovery.* London: Hutchinson.

Salmon, Wesley (1967) *The Foundations of Scientific Inference.* Pittsburgh: University of Pittsburgh Press.

Salmon, Wesley C. (1984) *Scientific Explanation and the Causal Structure of the World.* Princeton: Princeton University Press.

Savage, Leonard J. (1972) *The Foundations of Statistics.* New York: Dover.

Shafer, Glenn (1976) *A Mathematical Theory of Evidence*. Princeton: Princeton University Press.

Sober, Elliott (1991) *Reconstructing the Past: Parsimony, Evidence, and Inference*. Cambridge, MA: Bradford, MIT.

Stalker, Douglas, ed. (1994) *Grue! The New Riddle of Induction*. Chicago and La Salle, IL: Open Court.

Four Mistaken Theses
about Evidence, and
How to Correct Them

Peter Achinstein

A theory of evidence, as I understand that expression, provides conditions for the truth of claims of the form

(1) e is evidence that h,

where e is a sentence describing some state of affairs and h is a hypothesis for which e provides the putative evidence.[1]

Theories of evidence generally fall into one of two categories: subjective and objective. According to a subjective theory, a sentence of form (1) is true only if there is someone in a certain epistemic situation with respect to e, h, or their relationship. For example, on one type of subjective view, (1) is true for a particular person P only if the degree to which P believes h on the assumption that e is greater than the degree to which P believes h without the assumption that e. On a very different type of subjective view, a sentence of form (1) is true only if, when h was first formulated by some person or group, that person or group did not know whether e was true, and h was not first formulated with the intention of explaining or deriving e.

The former view is held by subjective Bayesians, who associate degrees of belief with probabilities, and require only that one's total set of subjective probabilities be "coherent," that is, satisfy the standard axioms of mathematical probability. For a subjective Bayesian

(2) e is evidence that h for person P at time t if and only if
$$p_{Pt}(h/e) > p_{Pt}(h),$$

that is, if and only if the subjective conditional probability of P for h on e at time t is greater than the prior probability of P for h at t.

The second subjective view is associated with William Whewell's doctrine of "consilience," according to which observational results that follow from a theory, but did not prompt the inventor(s) of the theory to formulate that theory, provide more evidential support for that theory than do observational results that prompted the theory in the first place.[2] On Whewell's view, whether e is evidence that h, and how strong that evidence is, depends at least in part on the epistemic situation of the person(s) who first formulated h.

According to the contrasting objective view of evidence, the truth of a sentence of form (1) does not depend on whether any person or group happens to be in some particular epistemic situation with respect to e, h, or their relationship. On an objective theory, indeed, a sentence of form (1) can be true even if no one knows or believes anything about e, h, or their relationship—indeed, even if there are no persons. On such a view, the concept of evidence is like the concept of a sign, or symptom, or indication, in at least one important sense of these terms. A change in atmospheric pressure can be a sign or indication of an impending weather change, even if no one knows about the change in pressure or knows that this is a sign or indication of anything.

Let me mention two prominent objective views: the hypothetico-deductive (h-d) view and Carnap's. According to a simple version of the former,

(3) e, if true, is evidence that h if and only if h entails e.

On this view, a fact described by a sentence e is evidence that h if and only if the sentence e is derivable from the hypothesis h. This can be so whether or not anyone knows or believes that e is true, and whether or not anyone knows or believes that e is derivable from h.

Carnap has an objective Bayesian view, according to which

(4) e is evidence that h if and only if $p(h/e) > p(h)$.[3]

where the probability in question is what he calls "logical" probability. He regards a conditional probability $p(h/e)$ as analogous to, and an extension of, the concept of entailment or implication in deductive logic. Carnap calls it "partial entailment." It represents the degree to which e entails h, where this is measured by real numbers between 0 and 1. Whether a sentence of the form $p(h/e) = r$ is true is an objective fact that does not depend on the epistemic situation of any person or group.

Carnap does provide an epistemic interpretation for his probability statements, but this too is objective. Consider a type of epistemic situation in which e expresses one's total knowledge of the results of one's observations. If $p(h/e) = r$, then a person in such an epistemic situation is justified in believing h to the degree r. This is objective because the truth of $p(h/e) = r$ does not depend on any person's being in an epistemic situation of the relevant sort.

A crucial difference between the subjective Bayesian view (2) and Carnap's objective Bayesian view (4) is that for the former but not the latter a sentence of the form "e is evidence that h" is true only for some particular person whose degree of belief in h on e is greater than his prior degree of belief in h.

37

In what follows, I consider four theses about evidence statements of type (1), particularly such statements in the sciences. The first is held by most objectivists as well as subjectivists. The next three are held by most objectivists. I suggest that each thesis is false and should be rejected. In this way I provide a basis for understanding some of the standard philosophical theories in the field. It also permits me to formulate briefly a theory of my own, which rejects each of the four theses to be discussed.

THESIS 1

Thesis

Evidential claims of form (1) made by scientists should be understood as being of one type. On a subjective view, these will be subjective evidential claims; on an objective view, they will be objective.

For subjective Bayesians, for example, all scientific evidential claims of form (1) should be understood as relativized to a particular person or group; they are claims about what such a person or group believes. For an objective hypothetico-deductivist, all scientific evidential claims of form (1) should be understood as claims about an objective deductive relation between an hypothesis and sentences reporting the putative evidence.

Response

Both subjective and objective evidential claims of form (1) are made by scientists all the time, and should be. Let me invoke an example to illustrate this. In 1883 Heinrich Hertz conducted experiments to determine whether cathode rays are electrical. He produced these rays in a cathode tube, separated them from the ordinary electric current that flows in the tube, and directed the rays into an electrometer which would determine the presence of electricity. The needle of the electrometer remained at rest. In another experiment he sent the cathode rays between oppositely electrified plates, which should deflect the rays if the latter are electrical in nature. No deflection occurred. Hertz concluded that

> (5) The results of these two experiments are evidence that cathode rays are electrically neutral.

Two theories had been proposed about the nature of cathode rays. Several German physicists, including Goldstein, Wiedemann, and Hertz thought they were electrically neutral waves of some type. Several British physicists, including Crookes and Schuster, believed they were charged particles—atoms or molecules of the gas in the cathode tube that have been negatively charged. In 1897 J. J. Thomson took up the question. He repeated Hertz's experiments, initially obtaining the same results. But, he reasoned, if the cathode rays really are electrically charged, then when they pass through the gas in the cathode tube they will ionize the gas molecules and produce positive and negative charges that in the parallel plates experiment will neutralize the charge on the plates, thus preventing deflection of the cathode rays. So if the gas in the tube is not sufficiently evacuated, there will be no deflection of the cathode rays. Thomson then proceeded to obtain a much higher degree of evacuation of the gas than Hertz was able to do 14 years earlier. When this was achieved, electrical deflection of the cathode rays was produced. Thomson concluded that cathode rays are charged particles rather than neutral waves, and in his experiments he was able to obtain a numerical value for the ratio of mass to charge of these particles (later called electrons).[4]

Is statement (5) true or false? Hertz claimed it is true, Thomson that it is false. Hertz's claim is based on the fact that he carefully designed the first experiment so as to separate the ordinary electric current from the cathode rays in such a way that (a) only the latter entered the electrometer; (b) the electrometer was working and was sufficiently sensitive to measure the electric charge, if any, carried by the cathode rays; (c) in the second experiment, since opposite charges attract, if the cathode rays were electrically charged they would be attracted by, and hence deflected toward, an oppositely charged electrical plate; and so on.[5] Thomson claimed that (5) is false because the experiments in question had a serious flaw: The gas in the tubes was not sufficiently evacuated to show electrical effects. This sort of disagreement over whether claim (5) is true—a disagreement in which one defends or criticizes (5) by appeal only to empirical facts pertaining to Hertz's experiments and not by an appeal to either Hertz's or Thomson's epistemic situation—involves an objective interpretation of the evidential claim (5).

However, there is another way to understand (5), in accordance with which both Hertz and Thomson would agree that (5) is true, namely, as a subjective evidence claim. One of the claims Hertz was making is that the results of his experiments were *his* evidence that cathode rays are electrically neutral. Fourteen years later, when Thomson reported on Hertz's experiments he also claimed that the results of these experiments were Hertz's evidence that cathode rays are not electrically charged. This (historical) claim was important for Thomson to report, because he regarded Hertz's experiments as worth repeating and initially obtained the same results as Hertz, but then realized that the experiments had a serious design flaw. Prior to Thomson's claim that Hertz's claim (5), objectively construed, is false, Thomson made an historical claim about what Hertz took to be evidence. This is claim (5), construed subjectively and relativized to Hertz.

More generally, both objective and subjective evidence claims of form

(1) e is evidence that h

are made (or denied) by scientists. When a scientist presenting his data e seeks to make an objective claim of form (1) he is also making a subjective one. He is claiming that e is his evidence that h, or what he takes to be evidence that h (in the objective sense). And when a scientist seeks to reject an objective evidence claim of some other scientist, he can do so by claiming that (1) is false if objectively construed, but true if subjectively construed and relativized to the scientist defending it. In rejecting an objective evidence claim of form (1), a scientist needs a way of stating whose evidence claim he is rejecting. This can be done using a subjective evidence claim.

THESIS 2

Thesis

In objective evidence claims of the form "e is evidence that h," the evidential relation is a priori not empirical: whether e, if true, is evidence that h is a matter to be determined by a priori calculation, not empirical investigation.

This thesis is held by defenders of a number of different objective theories of evidence. For example, on the simple h-d view, whether e, if true, is evidence that h depends simply on whether h entails e, which is an a priori matter. Even more sophisticated h-d views, which besides invoking "entailment" appeal to ideas about simplicity or coherence, are a priori, since whether these additional criteria are satisfied is supposed to be settleable without empirical investigation. On Carnap's view, according to which e is evidence that h if and only if $p(h/e) > p(h)$, the evidence relationship is also a priori, since for him, probability claims are to be understood in what he calls the "logical sense." Whether a probability claim is true is determined by a priori calculation.[6]

Not all objectivists subscribe to this a priori view, but many do.[7]

Response

40

If thesis 2 is understood as describing how scientists actually defend or criticize evidential claims, it is false. Consider again one of Hertz's objective evidential claims: *The fact that there was no deflection of the cathode rays by an electric field in Hertz's experiments is evidence that cathode rays are electrically neutral.* Thomson criticized this assertion on empirical, not a priori, grounds. He made the empirical claim that Hertz's cathode tubes were not sufficiently evacuated, and that when they were electrical deflection would be produced. And he actually carried out the experiment by producing a better vacuum in the tubes and achieving the predicted deflection. Thomson then made the claim that

> (6) The fact that there was deflection of the cathode rays by an electric field in his experiments is evidence that cathode rays are electrically charged.

This evidential claim can be defended in various ways, including by an appeal to the empirical facts (a) that opposite charges attract; (b) that without the electrically charged plates the rays are not deflected, but with them they are; (c) that nothing else in the tube is deflecting the rays, and if so, the only cause of deflection is the electric field produced by the plates; and (d) that the cathode rays will be deflected by the plates only if the rays carry an electric charge. Thomson's evidential claim (6) was not, and cannot be, defended on a priori grounds, simply by mathematical or logical calculations.

An a priorist defender of Thesis 2 may reply by admitting that evidential claims such as (6), as usually formulated, are empirical. But he may insist that they should be rendered a priori by incorporating into them all of the empirical information one might appeal to in their defense. So if the original claim is of the form

> (1) e is evidence that h,

and if (1) is *empirically establishable in the most complete way possible* by appeal to empirical background information b, then claim (1) should be reformulated as

> (7) e is evidence that h, given b.

We are to understand the claim that (1) is "empirically establishable in the most complete way possible by appeal to b" as entailing that the sentence of form (7) is a priori.

I do not want to deny that one might be able to pack enough empirical information into b to make (7) a priori. That would not show that (1) is a priori. Nor would it show why it is required or even advantageous to assert an a priori version of (7) rather than an empirical version of (1), or of (7). Perhaps transforming an empirical claim into an a priori one that incorporates a complete empirical defense is one way of establishing the truth or falsity of the empirical claim. But it is not the only way, or necessarily the best way. Thomson discovered the falsity of Hertz's evidential claim (5), and the truth of his own evidential claim (6), not by turning the former into a false a priori claim and the latter into a true one, but by conducting the experiments he did.

Moreover, if it is agreed that a claim such as (6) is empirical, then, as with any empirical claim, whether evidential or not, why should one attempt to transform it into an a priori one by relativizing it to its most complete empirical justification? Empirical claims should be treated as such. One can separate the empirical claim itself from any justification one might offer for it, whether or not relativizing the claim to that justification yields an a priori claim.

THESIS 3

Thesis

Objective evidential claims must always be understood as relativized to an epistemic situation.

Objective evidential claims of the form "e is evidence that h" must be understood as saying that e is evidence that h *for anyone in an epistemic situation ES in which certain beliefs are held.*[8] There need not be anyone in the epistemic situation to which the evidential claim is relativized. Nevertheless, the claim is to be understood as relativized to a (type of) situation.

Consider again Hertz's evidential claim

(5) The results of these (Hertz's) two experiments are evidence that cathode rays are electrically neutral.

We might consider this claim relativized to an epistemic situation containing the beliefs of Hertz in 1883. If we do, then claim (5) can be regarded as true in the sense that anyone, including Hertz, in such an epistemic situation would be justified in believing that cathode rays are neutral on the basis of the experiments Hertz conducted. However, we might also consider the claim (5) relativized to an epistemic situation of the sort Thomson was in 14 years later, in 1897, which includes the results of Thomson's experiments. Anyone, including Thomson, in that situation would not be justified in believing that cathode rays are neutral on the basis of the experiments that Hertz conducted in 1883. Anyone in that epistemic situation would be justified in believing that Hertz's experimental design contained a flaw.

Without such relativizations evidential claims have no truth-value.

Response

42

I agree that there is a concept of objective evidence that is relativized to an epistemic situation (I call this ES-evidence). With this concept we can indeed say that claim (5) is true: Relative to an epistemic situation of the sort Hertz was in, the results of his experiments were evidence that cathode rays are electrically neutral.

However, in addition to such a relativized concept of evidence, there is also a non-relativized one. Using a non-relativized concept, we can say that since Hertz's experiments were experimentally flawed because the cathode tubes were not sufficiently evacuated, the results of these experiments are not evidence that cathode rays are neutral. Earlier I noted an analogy between evidence and the concept of a sign. We speak of signs in two ways: (1) relativized: a sign for someone in a certain epistemic situation (these sorts of spots are a sign of the presence of the measles virus for physicians today, but not during the nineteenth century); and (2) non-relativized (these sorts of spots are a sign of the presence of the measles virus, irrespective of any epistemic situation).

In the non-relativized sense, to determine whether X is a sign of Y, we must determine (roughly) whether when X is present Y is also usually present. In the relativized sense, to determine whether X is a sign of Y, we don't need to determine this. Instead we need to identify an epistemic situation—either a type or a situation that someone is actually in. If it is a type, we can say that X is a sign of Y for anyone in that situation if anyone in that situation would be justified in believing that X is a sign of Y (in the non-relativized sense). If it is the epistemic situation of some particular person, we can also say this, or we can say (even more subjectively) that X is a sign of Y for some person who believes that X is a sign of Y (in the non-relativized sense), whether or not that person is justified in so believing.

Similarly, I claim, there is not only a relativized concept of evidence but a non-relativized one as well. In the latter case, if e is evidence that h it is so independently of epistemic situations and of persons or groups in such situations. To be sure, when Thomson made the evidential claim (6), he can and did defend it by appeal to facts he believed, in his epistemic situation. But that does not mean that his evidential claim must be understood as relativized to what he believed. Otherwise why shouldn't every empirical claim one makes, for example, the claim that cathode rays are charged, be so relativized?

There is a distinction between (a) a relativized statement such as "given what Thomson believed in 1897, he was justified in believing that cathode rays are charged;" and (b) an evidential claim, such as Thomson's (6), that is *defended* by appeal to what Thomson believed in 1897. In case (a) the claim itself ("given what . . .") is relativized. In case (b) the evidential claim (6) is not relativized to what Thomson believed in 1897.

43

THESIS 4

Thesis

Objective evidence is a weak idea. You don't need very much to have evidence for a hypothesis.

This thesis is best explained by looking briefly at two well-known objective theories.

> *Simple hypothetico-deductivism:* e if true is evidence that h if and only if h entails e.

This entailment condition is very weak. Many facts are entailed by a hypothesis, all of which would have to count as evidence that the hypothesis is true. For example, the fact that the sun exists is entailed by Kepler's first law that the planets revolve around the sun in elliptical orbits. So the fact that the sun exists would be evidence that Kepler's first law is true. The fact that I own at least one ticket in a million-ticket lottery is entailed by the hypothesis that I will win the lottery, together with the background information that in order to win you must own at least one ticket. So, given this background information, the fact that I own at least one ticket is evidence that I will win. This, it must be concluded, is a very weak concept of evidence.[9]

> *Increase-in-probability view (positive relevance):* e is evidence that h if and only if $p(h/e) > p(h)$; or, e is evidence that h, given b, if and only if $p(h/e\&b) > p(h/b)$.

Again, this is a very weak condition, since all that is required to obtain evidence that h is to find something that increases h's probability. So, for example, the fact that I buy one ticket in a million-ticket lottery counts as evidence that I will win, since it increases the probability that I will (from 0 to 1/1 million). The fact that I am entering an elevator counts as evidence that I will be in an elevator accident, since it increases the probability that I will. Such examples could be multiplied endlessly. The positive relevance concept demands very little of evidence.

A number of other objective views, such as Hempel's satisfaction idea (which is a version of inductivism)[10] and Glymour's bootstrapping theory,[11] make similarly weak demands on evidence. On these views it takes very little to obtain evidence.

Response

Scientists and others want evidence that h because evidence that h gives a good reason to believe h. If so, then evidence cannot be a weak concept, one that allows the sorts of cases cited above. In these cases the facts invoked do not provide a good reason to believe the hypotheses in question. The fact that the sun exists is not (at least not by itself) a good reason to believe that the planets revolve about the sun in elliptical orbits, despite the fact that the latter entails the former. The fact that I am entering an elevator is not (by itself) a good reason to believe that I will be in an elevator accident, despite the fact that it increases the risk.[12]

In the next section I briefly develop a much stronger concept of evidence (or rather several concepts), in violation of each of the four theses I have been discussing.

STRONG EVIDENCE

Following the positive relevance theorists, I believe that evidence is related to probability, but that increase in probability is neither necessary nor sufficient. On my view,

> (a) For a hypothesis h and putative evidence e, if e is a good reason to believe h, then there is some number k (greater than or equal to 0) such that $p(h/e) > k$.

> (b) If e is evidence that h, then e must be a good reason to believe h.

> (c) If e is a good reason to believe h, then e cannot be a good reason to believe not-h.

Assumptions (a)–(c) can be shown to require that k in (a) be $1/2$,[13] so that

> (d) e is evidence that h only if $p(h/e) > 1/2$.

In accordance with (d), the fact that I am tossing this fair coin (e) is not evidence that it will land heads (h), since $p(h/e) = 1/2$. It is not a good reason to believe that the coin will land heads, because it is an equally good reason to believe that it won't. But, in accordance with (c), if e is a good reason to believe h, then it cannot be a good reason to believe not-h. (The fact that I am tossing this fair coin is just as good a reason for believing it will land heads as that it will land tails, but that doesn't make it a good reason; two equally good reasons can both be quite bad.)

If (d) is a necessary condition for evidence, as I believe it to be, then the positive relevance view fails to provide a sufficient condition for evidence. This is the case, since it can be that $p(h/e) > p(h)$, even though $p(h/e) < 1/2$. For example, the fact that I own one ticket in a fair lottery with 1000 tickets is not evidence that I will win, since $p(h/e) = 1/1000 < 1/2$, even though $p(h/e) > p(h)$. Similarly, if we accept (d), then the hypothetico-deductive view also fails to provide a sufficient condition for evidence, since h can entail e, even though $p(h/e) < 1/2$. For example, let h be that I own all the tickets in the lottery, and let e be that I own the ticket marked "1". The fact that e is true, assuming it is, is not evidence that h.

45

In these examples e is not evidence that h, since e does not provide a good reason for believing h, which itself requires that $p(h/e) > p(\text{not-}h/e)$, in other words, that $p(h/e) > 1/2$. It can also be shown that the positive relevance and h-d views fail to provide necessary conditions for evidence, that is, e can be evidence that h even though $p(h/e)$ is less than or equal to $p(h)$, and even though h does not entail e.[14]

More importantly, although condition (d) is necessary for evidence it is not sufficient. The probability of h can be high irrespective of e. For example, let

e = Michael Jordan eats Wheaties
h = Michael Jordan will not become pregnant

$p(h/e) = p(h)$, which is approximately equal to 1. Yet no one wants to say that the fact that Michael Jordan eats Wheaties is evidence that he will not become pregnant. The former fact is not a good reason for believing the latter, which I take to be necessary for evidence. Some further condition making e evidentially relevant for h is required.

To obtain this condition I suggest the existence of a relationship between evidence and explanation, namely,

(e) e is evidence that h only if the probability is high that there is an explanatory connection between h and e, given h and e. (I will take high probability as greater than $1/2$.)

There is an explanatory connection between h and e if and only if either h correctly explains why e is true; or e correctly explains why h is true; or some hypothesis correctly explains both why e is true and why h is true.[15] In the Michael Jordan example, the probability of an explanatory connection between h and e, given h and e, is very low. (Given that Michael Jordan will not become pregnant and that he eats Wheaties, the probability is very low that he will not become pregnant because he eats Wheaties, or that he eats Wheaties because he will not become pregnant, or that some hypothesis correctly explains both why he eats Wheaties and why he won't become pregnant.)

Although I take (e) to be a necessary condition, I argue that the following stronger condition is necessary, which conditionalizes the probability not on h and e, but on e alone:

> (f) e is evidence that h only if the probability is high that there is an explanatory connection between h and e, given e.

Condition (f) can be shown to hold if and only if the product of the probabilities in (d) and (e) is greater than $1/2$, that is, if and only if

> p(h/e) x p(there is an explanatory connection between h and e/h and e) > $1/2$.

To complete the requirements, I add that e must be true and that e must not logically entail h. These together with condition (f) comprise the definition. So we have

> (g) e is evidence that h if and only if
> - (i) p (there is an explanatory connection between h and e/e) > $1/2$. Equivalently, p(h/e) x p(there is an explanatory connection between h and e/h and e) > $1/2$;
> - (ii) e is true;
> - (iii) e does not entail h.

This, I claim, suffices to rule out the counterexamples to other views. It also yields an appropriate relationship between evidence and a good reason to believe (if e satisfies these conditions, then e is a good reason to believe h).

The concept of evidence given in (g), which I call *potential* evidence, does not require the truth of h for e to be evidence that h. A stronger concept, which I call *veridical* evidence, does:

> (h) e is veridical evidence that h if and only if
> - (i) e is potential evidence that h;
> - (ii) h is true;
> - (iii) there is an explanatory connection between e's being true and h's being true.

The weaker and stronger concepts of evidence reflected in (g) and (h) parallel weaker and stronger concepts of "sign" and "symptom." In the weaker sense, X (e.g., these spots) can be a sign or symptom of Y (e.g., the measles virus) even if it turns out that Y (the virus) was not present. In the stronger sense, if Y was not present then, although X may have been a sign or symptom of something, it was not a sign or symptom of Y.

Two other concepts of evidence can be characterized using veridical evidence. One is evidence *for anyone in a given epistemic situation,* which I call ES-evidence:

> (i) e is ES-evidence that h, relative to an epistemic situation ES, if and only if e is true, and anyone in ES is justified in believing that e is (probably) veridical evidence that h.

The second is subjective evidence for some person or group X:

47

> (j) e is X's subjective evidence that h at time t if and only if at t, X believes that e is (probably) veridical evidence that h, and X's reason for believing h true (or probable) is that e is true.

Of the four concepts of evidence characterized in (g), (h), (i), and (j), the most important, in the sense that it reflects what scientists seek, is veridical evidence. Scientists want their hypotheses to be true, and they want to provide a good reason for believing them in a sense of "good reason" that requires truth.[16] This is what veridical evidence supplies. However, the most fundamental concept is potential evidence, since the others are defined by reference to it.

All four concepts are employed in the sciences. When we claim that Thomson's deflection experiments provided evidence that cathode rays are electrically charged, our claim is that the results of these experiments are veridical, and hence, potential, evidence of the truth of this hypothesis. It is also true that these results were Thomson's subjective evidence that h (he believed that the results were veridical evidence of the electrical nature of cathode rays and his reason for believing is that that he obtained the deflection he did). Finally, we say that Thomson, and by implication anyone, in his epistemic situation, was justified in believing that these results constitute veridical evidence that cathode rays are charged. We claim that these results are ES-evidence that h for anyone in that epistemic situation.

Potential, veridical, and ES-evidence are objective concepts: e can be evidence that h in one of these senses without there being a person or group that actually believes e, or h, or that e is evidence that h. Only subjective evidence requires the existence of such a person or group. This means that the concept of probability used to define potential evidence is an objective rather than a subjective one.[17]

Four Theses Redux

Armed with these concepts of evidence, let us briefly return to the four rejected theses.

Thesis 1: Evidence claims of the form "e is evidence that h" should all be understood as being of one type, either subjective or objective.

Response: Both objective and subjective concepts of evidence are employed. Moreover, the various objective concepts I have introduced in earlier sections are in use, not just one.

Thesis 2: In objective evidence claims, whether e if true is evidence that h is a priori, not empirical.

Response: Although there are some a priori objective evidence claims, for the most part objective evidence claims are empirical. Veridical evidence claims involving an empirical hypothesis h are empirical since they require the truth of h. Potential evidence claims require the truth of sentences of the form p(there is an explanatory connection between h and e/e) > 1/2, which are usually defended empirically. (In Thomson's case, the probability claim would be that given the deflection of the cathode rays in the tube containing oppositely charged plates, the probability is high that the deflection is caused by the presence of an electric field, and hence the rays are charged. Such a claim would be defended by appeal to various empirical facts, including that opposite charges attract.)

An ES-evidence claim is usually empirical. Here is a simple example: *The deflection of cathode rays in Thomson's experiments is ES-evidence that cathode rays are charged, given Thomson's epistemic situation in 1897.* This is empirical, since it is an empirical question what Thomson's epistemic situation was in 1897. To be sure, one might enumerate all the contents of that epistemic situation, and relativize the evidence claim to that—which could result in an a priori claim. But this is rarely if ever done, nor does it need to be. To make an empirical ES-evidence claim such as the one above it is not required to produce, or even know, all or any of the contents of Thomson's epistemic situation that would be cited in defending the claim.

Thesis 3: Objective evidence claims must always be understood as relativized to an epistemic situation.

Response: This is true for ES-evidence claims, but not for ones involving potential or veridical evidence. The truth of claims of the latter types does not depend on any epistemic situation that is being assumed. In this respect such claims are like corresponding ones of the form "X is a sign, or symptom, of Y." Whether these spots are a sign or symptom of the measles virus does not depend on what beliefs are presupposed.

Thesis 4: Objective evidence is a weak idea. You don't need very much to have evidence for a hypothesis.

Response: The three objective concepts I have introduced are strong, not weak. With veridical evidence you have a good reason to believe a hypothesis in a sense that requires it to be true. With potential evidence you have a good reason to believe a hypothesis, in a sense that requires it to be more probable than not. With ES-evidence you are justified in believing that e is (probably) veridical evidence that h, and hence that h is (probably) true. These are strong, not weak, demands. They will not countenance numerous cases allowed by the much weaker concepts advocated by the positive relevance, h-d, and other such views.

49

In this chapter I have concentrated on theories that provide truth-conditions for sentences of the form "e is evidence that h." But there are many other issues about scientific evidence that can be raised. On a general level, for example, one might ask how to proceed to gather information e that will count as evidence for a hypothesis h.[18] On a more specific level, one might ask (especially in controversial cases) whether some particular body of information does or does not count as evidence for a given hypothesis, and why.[19] Philosophers who propose general theories of evidence of the sort I have criticized and defended hope that their theories will aid in solving these other issues as well.

NOTES

Ideas in this paper come from my *The Book of Evidence* (New York: Oxford University Press, 2001; hereafter BE). I repeat them here not only because they are in conflict with standard philosophical views of evidence, but also because several of the contributors to the present volume accept, reject, or modify my views in their essays. For very helpful suggestions I am indebted to Fred Kronz and Amy McLaughlin.

1. More accurately, it is the fact that e, rather than a sentence describing that fact, that constitutes the evidence. But here I follow the standard philosophical practice of speaking of the sentence e.

2. William Whewell, *Philosophy of the Inductive Sciences,* vol. 2 (New York: Johnson Reprint Corporation, 1967; reprinted from the 1847 edition).

3. Rudolf Carnap, *Logical Foundations of Probability* (Chicago: University of Chicago Press, 2nd ed., 1962). Carnap offers a second concept, for which high probability rather than increase in probability is necessary and sufficient: e is evidence that h if and only if $p(h/e) > k$, where k is some threshold for high probability.

4. For more details of the story see BE, ch. 2.

5. For a defense of Hertz's reasoning, see Jed Z. Buchwald, *The Creation of Scientific Effects* (Chicago: University of Chicago Press, 1994), pp. 150–74.

6. For a simple exposition of Carnap's theory of probability see BE, pp. 49–50.

7. There are some prominent exceptions, such as the error statistical account of Deborah Mayo, *Error and the Growth of Experimental Knowledge* (Chicago: University of Chicago Press, 1996), and the material theory of induction of John Norton. See the chapters by these authors in this volume.

8. See Kent Staley's chapter in this volume for a version of this thesis.

9. There are attempts to strengthen the hypothetico-deductive idea by requiring further conditions, for example, that e be a prediction and not already known to be true, or that h be, or be part of, a coherent system of hypotheses. Whether these additions sufficiently strengthen the concept is controversial. In *Particles and Waves* (New York: Oxford University Press, 1991), I argue that these additions are neither necessary nor sufficient.

10. Carl G. Hempel, *Aspects of Scientific Explanation* (New York: Free Press, 1965).

11. Clark Glymour, *Theory and Evidence* (Princeton: Princeton University Press, 1980).

12. For a defense of weaker concepts of evidence against my criticisms, see Steven Gimbel's chapter in this volume.

13. For the argument see BE, pp. 115–16.

14. For examples see BE, pp. 69–71, 147. Sherrilyn Roush's chapter in this volume rejects my argument that positive relevance is not necessary for evidence; my reply is included in the second part of that chapter.

15. For a definition of "correctly explains" see BE, pp. 160–66.

16. In BE, chapter 8, I defend this "realist" conception of evidence. Scientific realism itself is a doctrine I defend in "Is there a Valid Experimental Argument for Scientific Realism?" *Journal of Philosophy*, vol. 99 (Sept. 2002), pp. 470–95. For a defense of a more modest realism, see Laura Snyder's chapter in this volume.

17. An account of such a concept, different from standard ones, is given in BE, chapter 5.

18. This is a question raised by Kronz and McLaughlin in their chapter in this volume.

19. For example, in this volume Richard Richards asks whether domestic breeding provides evidence for or against Darwin's theory of evolution.

Restoring Ambiguity to Achinstein's Account of Evidence

Steven Gimbel

Thresholds and the Denial of Ambiguity

In his latest work, *The Book of Evidence*, Peter Achinstein argues that *evidence* is ambiguous in a number of ways. We may legitimately speak of evidence in both an objective and a subjective, biographical sense and we may speak of evidence relativized to an epistemic situation and of evidence which is not so relativized. But he rules out the ambiguity between supporting and confirming evidence, between evidence-for and evidence-that; in other words, between facts in the world that advance the case of a hypothesis and facts in the world that provide good reason to believe a hypothesis. Since evidence, in his view, must provide nothing short of good reason to believe, a notion of increased epistemic support that applies to hypotheses for which there is not legitimate, objective grounds for belief is inherently and unrevisedly problematic.

However, the new twist on Achinstein's long-standing view in *The Book of Evidence* does not, in fact, exclude supporting evidence; rather, his new framework for the explication of evidence provides a quite natural picture of evidence-for, which is supported by no less than Achinstein's own earlier historical case studies. When we examine his discussions of Maxwell's early work on the mechanical theory of heat, Bohr's quantization of the hydrogen atom, and Perrin's argument for the objective reality of atoms, a robust picture of empirically based argumentation in the

The text of this chapter was previously published in the *British Journal for the Philosophy of Science*, 55 (2004): pp. 269–85.

context of pursuit emerges. This picture may be married to Achinstein's notion of "evidence as a threshold concept," which then gives rise to an account of supporting evidence that, contrary to his own argumentation, is entirely consistent with and complimentary to his new explication of confirming evidence.

ACHINSTEIN'S NEW ACCOUNT OF
CONFIRMING EVIDENCE

The view set out in *The Book of Evidence*, with roots in Achinstein's writings from the past 25 years, remains a synthesis of probability and explanatory models. But Achinstein moves beyond his earlier arguments, providing several novel twists on his long-standing view. Key to the view is a revision of the relation between rational belief and probability of truth.[1] Whereas Bayesians, on the one hand, identify the two, and some like Mark Kaplan (1996), on the other hand, divorce rational belief from posterior probability, Achinstein posits a more intricate relation, taking evidence to be what he calls a "threshold concept."

> [C]oncepts such as acceptability, having some foundation, firmness, and confidence, in so far as they depend on probability, are "threshold" concepts. A necessary condition that must be satisfied for a hypothesis *h* to have any acceptability, foundation, or firmness, and before I have any confidence in it, is that *h*'s probability exceed some threshold. In this regard, these concepts (with respect to probability) are like the following concepts (with respect to the items in parentheses): crowds (with respect to the number of people), electrically induced pain (with respect to voltage), and my fear of death due to playing a sport (with respect to the rate of death in that sport). Each of these concepts is related to the concepts in parentheses. But there needs to be a certain threshold amount falling under the concept in parentheses in order that any amount of the concept dependent on it is realized. One person in the audience is not a crowd, even a small one. Two volts of electricity are not sufficient to produce any pain in me. A death rate of 1 per 500 million rounds of golf (as reported in the *New York Times Magazine* June 18, 1995) is not enough to produce any fear in me that I will die playing golf. (Achinstein, 2001: 73–74)

Thresholds induce a discontinuous structure upon a seemingly smooth concept.

The notion of a threshold appears at two critical junctions in Achinstein's account. First, Achinstein separates reasonableness of belief from probability of truth. Consider a proposition with very small, but non-zero probability. Just as one person is not a small crowd, so "[a] probability of death from play-ing a round of golf equal to 1/500 million is not sufficient to provide good reason to believe, or a justification for my believing, that I will die during my next round of golf" (Achinstein, 2001: 74). Absurd hypotheses are not those for which we have a small degree of belief; rather, they are those whose prob-ability is so low as to provide *no* reason for rational belief.

But if we have no belief in absurd hypotheses, then facts[2] that make an absurd hypothesis only slightly more likely so that it still fails to be moved beyond the minimum threshold of rational belief are—and should be—regarded as epistemologically irrelevant.

> [A]ssuming I buy 1 ticket in a 100 million ticket lottery, where one ticket will be selected at random to win, I increase the probability that I will win from 0 to 1/100 million. However, I increase my confidence in winning (firmness of the support of the hypothesis that I will win, etc.) only if I come to have at least some confidence in winning (the hypothesis has some firmness of support, etc.). But I have no such confidence (the hypothesis has no firmness of support). There is not enough probability here to exceed a threshold necessary for me to have any confidence (for the hypothesis to have any firmness, support, etc.). (Achinstein, 2001: 74)

Contrary to the standard conception, Achinstein argues that some changes in posterior probability warrant absolutely no rational confidence at all in the hypothesis, and are not evidence.

The relationship between reasonableness of belief and probability, Achinstein argues, may be represented graphically as a step function. For a range of probability, the reasonable level of acceptability of the hypothesis will remain constant. But at—and only at—crucial points, belief in a hypothesis exhibits a "quantum jump" to the next "energy level" of acceptability. The first sense in which evidence is a threshold concept, then, is that reasonableness of belief bears a threshold relationship to probability of truth.

The task becomes locating epistemological thresholds that are objec-tively definable. It is here that we find the second sense in which evidence is a threshold concept. The one threshold that is epistemically relevant to evidence is that for which *e provides at least some good reason to believe h*. Any weaker condition runs into trouble, Achinstein argues, because if we use the term *evidence* for a fact that does not provide *good* reason, but only *some* reason to believe a hypothesis, then it may allow a single fact to stand as evidence for both a hypothesis and its negation. This does not make it evi-dence both for and against the hypothesis, but rather no evidence at all. That a fair coin is being flipped is *some* reason to believe that it will land heads and some reason to believe that it will not, but it is not *good* reason to believe either. Evidence is not like force, but like net force. It would be better to say that the flipping of the coin does not provide any evidence that the coin will land heads up than to say that it provides evidence that it will or that it will not land heads up.

The latter option gives rise to a very impoverished notion of evidence, and this weakness ought to be avoided. The success of any philosophical account of scientific evidence, Achinstein argues, must hinge upon its usefulness to the living practice of science: "[S]tandard philosophical theories about evidence are (and ought to be) ignored by scientists. They ought to be ignored because they propose concepts of evidence that are based on assumptions incompatible with ones scientists make when they speak of, and offer, evidence for hypotheses" (Achinstein, 2001: 3). The usual accounts of evidence, Achinstein argues, are rightfully ignored by scientists because they are too weak to provide the one thing scientists want from evidence—good reason to believe.

54

An explication of *good reason to believe* must require some sort of high probability condition relating e to h, but to avoid the problem of irrelevant information[3] we must include both the high posterior probability of h on e and the high probability of an explanatory connection between h and e, given h and e.[4] Earlier forms of the model (see 1983*a* and *b*) call for the conjunction of the high probability conditions,[5] but in the new model it is not the individual probabilities, but their product, that must surpass the .5 threshold.[6] This multiplicative definition is significantly stronger than the old conjunctive one. Facts that just squeaked in under the conjunctive definition would have a multiplicative value close to .25, and would therefore fail to meet the new condition by far.

One interesting result that Achinstein derives in the name of parsimonious presentation is the equivalence of the multiplicative condition with a condition involving only the explanatory connection. Since the existence of an actual explanatory connection given h and e entails the truth of the hypothesis, it can be proven that Achinstein's multiplicative definition is satisfied if, and only if, the probability of the existence of an explanatory connection given only the truth of e is high, that is, $p(h/e)p(expl[h,e]/h,e) > 1/2$ is equivalent to $p(expl[h,e]/e) > 1/2$ (Achinstein, 2001: 153–54).

ACHINSTEIN'S ARGUMENT AGAINST THE AMBIGUITY RESPONSE

Based on this view, Achinstein argues that he has defeated what he terms the *ambiguity response*, the claim that there are two distinct notions of evidence: supporting evidence and confirming evidence. The view goes back at least as far as Carnap (1962: xv–xx) and was set out clearly by Salmon:

> As Carnap pointed out in *Logical Foundations of Probability*, the concept of confirmation is radically ambiguous. If we say, for example, that the special theory of relativity has been confirmed by experimental evidence, we may have either of two quite distinct meanings in mind. On the one hand, we may intend to say that the special theory has become an accepted part of scientific knowledge and that it is very nearly certain in light of its supporting evidence. If we admit that scientific hypotheses can have numerical degrees of confirmation, the sentence, on this construal, says that the degree of confirmation of the special theory of relativity on the available evidence is high. On the other hand, the same sentence might be used to make a very different statement. It might be taken to mean that some particular evidence—e.g., observations on the lifetimes of mesons—renders the special theory more acceptable or better founded than it was in the absence of this evidence. If numerical degrees of confirmation are again admitted, this latter construal of the sentence amounts to the claim that the special theory has a higher degree of confirmation on the basis of the new evidence than it had on the basis of the previous evidence alone. (Salmon, 1975: 5)

Scientists, the line goes, use *evidence* to mean both good reason and more reason to believe.

On this view, Achinstein's explication of good reason to believe would be at best satisfactory for only one sense of scientific evidence, making his view incomplete. Some account of increased epistemic support would be needed to supplement his account of good reason to believe. But Achinstein objects to this charge of incompleteness on the grounds that any attempt to formulate an account of "more reason to believe" will yield a concept that is doomed to fail in at least two ways.

The first is the absurd hypothesis scenario discussed above. In cases of highly improbable hypotheses, Achinstein argues, advocates of supporting evidence will be forced to attribute the title *evidence* to facts that inspire no change in our complete lack of rational belief in absurd hypotheses. Since increase in posterior probability does not necessarily entail any rational confidence whatsoever, much less an increase in rational confidence, the concept of evidence-for should be abandoned. He states, "I reject the ambiguity response. Even if probabilists were correct in supposing that there is a sense of evidence that involves the idea of increase-in-strength-of-evidence, and even if the latter is connected to probability, it does not follow, and indeed is false, that any increase in probability is an increase in the strength of the evidence" (Achinstein, 2001: 74).

The second objection that Achinstein levels at the concept of supporting evidence is that it necessarily runs afoul of his pragmatic concern. The notion of supporting evidence will necessarily be weaker than his very robust picture of confirming evidence, and will thereby be insufficient for the practitioners of science. Such accounts will be forced—for example, to say that a perfectly healthy Olympic swimmer's entry into the pool for morning laps is *some* evidence-for the hypothesis that the swimmer will drown (Achinstein, 2001: 70)—and such results are the reason that scientists legitimately heap scorn on the standard philosophical attempts to explicate the notion of scientific evidence. The only scientific context for which it makes sense to speak of a useful and objective sense of evidence is confirming evidence that provides good reason for belief.

Supporting evidence is thus problematic, in that (1) it cannot account for evidence as a threshold concept and (2) it is too weak to provide a useful account for working scientists. Achinstein considers both of these concerns to be insurmountable, and therefore the pursuit of an explication of supporting evidence is pointless. Accordingly, the claim of the ambiguity is not well grounded, and the charge of incompleteness is unfounded.

A Threshold-based Approach for Restoring the Ambiguity

These concerns are not unanswerable. It is possible to build an account of supporting evidence that accounts for both of Achinstein's objections and fits the more general constraints that Achinstein places on a successful account of evidence. Indeed, we may find both the broad-stroke blueprint and details of such an account of evidence-for in Achinstein's own writings.

Achinstein's first objection unnecessarily singles out Carnap's explication of the concept of supporting evidence, instead of approaching the notion of evidence-for itself. If Achinstein is correct about rational belief, and derivatively, about evidence being a threshold concept—let us assume that he is—then there is no doubt that the positive relevance view misses the mark. But Carnap's failure does not imply that there is not a view that can avoid the pitfall. Let us then assume the stepwise relation of reasonable belief to a combination of probabilistic and explanatory relations as a starting point. The question then becomes whether there is a threshold-friendly picture of supporting evidence that meets Achinstein's other criteria.

Regarding Achinstein's second concern, there is no doubt that scientists do talk in terms of evidence-that, but it is not clear why Achinstein's pragmatic concern can only be satisfied through an extremely strong notion of confirming evidence. Scientists certainly do want good reason to believe hypotheses and they do want evidence to provide this, but that is not the only use to which they put it. There is a threshold-based notion of evidence-for that can be shown to be germane to actual scientific practices.

But to satisfy Achinstein, we require not only a threshold-based picture of evidence-for that can be shown to be scientifically relevant, but also one that is (1) *objective* in the sense that whether or not *e* is evidence for hypothesis *h* is not a matter of what anyone believes or does not believe, and (2) *empirical* in the sense that *e* being evidence for *h* is not a matter demonstrable by a priori calculation. This goal may be achieved if we can locate objective thresholds weaker than "good reason to believe," namely, other steps in the epistemological step function that are scientifically relevant. The existence of such thresholds would reinstate the ambiguity.

A case for such thresholds can be made from Achinstein's own writings. There is no doubt that scientists ultimately aim to show that there is or is not good reason to believe. But Achinstein points out that at times in the scientific process,

> the aim is to show that the theory is *reasonable to pursue*; i.e., that it is reasonable to try to work out consequences of the theory, to apply it to more complex systems, add or reformulate assumptions, devise possible ways to test it, and so forth. Doing this is important because when a theory is first proposed the scientific community needs to know whether research on it should be encouraged, and whether tests should be planned and conducted. Funding agencies with limited resources need to know whether proposals to investigate the theory should be supported. And scientists need to know whether it is reasonable to invest their own time and energy in its pursuit. (Achinstein, 1993: 90)

So here we have scientific reasoning focused on something other than providing good reason to believe, that is, a part of the scientific process where rational arguments are based on empirical statements in order to exceed a less stringent epistemic threshold. Understanding argumentation in the context of pursuit may offer exactly the sort of threshold we are looking for.

One concern is that the sort of evidence sought by scientists is wholly non-contextual by Achinstein's lights; if *e* is evidence with respect to *h*, then *e* is good reason to believe *h*, independent of what else anyone knows or believes. But a hypothesis' rational worthiness of pursuit is ineliminably context-dependent, because the notion of rational pursuit is teleological and there is any number of aims which might be served by pursuing a given hypothesis. Let us bracket all of these except the acquisition of propositions that we have good reason to believe—in other words, let us consider only that which is relevant to the epistemic reasons for concern and not any of the myriad of non-epistemic reasons one might have for pursuing a hypothesis. Deep-context dependence still remains.

First and foremost, reasonableness of pursuit is historically bound. It was perfectly reasonable in 1800 for Thomas Young to pursue the wave theory in light of refraction phenomena, or for Joseph Priestly to pursue phlogiston theory, given the calcination of metals, but these facts could not be said to make it reasonable to pursue these theories in 2000. What we come to know affects the reasonableness of pursuit, making the inference deeply pragmatic.

Further contextual considerations in the service of acquiring propositions for which we have good reasons for belief include resource availability and technological state. The general theory of relativity, for example, sat dusty on the shelf for decades because the computation of the components of the metric tensor required calculations that are so time-consuming for any non-trivial mass–energy distribution. It was not until the introduction of computer programs capable of shouldering much of the labor that working with the theory in certain interesting ways became feasible. Further, once results that deviated from classical theory were derived, the ability to construct instrumentation sophisticated and sensitive enough to collect the relevant data had to be waited for, which then brought concerns of funding such projects into play. Reasonableness of pursuit is therefore sensitive to background knowledge, technological advancement, and resource allocation.

Political considerations are also relevant. The decisions of Galileo and Bruno to pursue the Copernican model involved more than considerations of the fruitfulness of the program. Only slightly less grave are the ramifications for junior faculty in selecting a research project. A trendy, but scientifically less promising option may be more reasonable to pursue when, say, securing grant money or the production of publishable results in short order are imperative. So it is clear that reasonableness of pursuit is not in and of itself the sort of objective criterion that is sought.

But the fact that the inference governing the reasonableness of pursuit is essentially pragmatic does not imply that there is no objective evidential relation residing within such inferences. Consider what happens when we bracket the pragmatic concerns. For example, if a colleague opts to work in intelligent design theory in order to keep a hard-won position at a religious-affiliated college where his spouse is also teaching, we recognize and are often emotionally affected by the need to make that choice. We may not deny that he made the rational choice for someone *in his circumstances*, but the emotional distress one feels about the admittedly rational choice clearly points to the fact that there is something underneath the pragmatic concerns other than more merely pragmatic concerns.

The fact that he would not be pursuing this project, all other things being equal, gives us the hint that there is something epistemologically grounded in our viewing of the different research options apart from the context. In other words, what serves as the foundation for the *ceteris paribus* claim of reasonableness is independent of the contextual facets bracketed, since we can continue playing the bracketing game without changing our ultimate assessment that the option not pursued is preferable. Thus there seem to be non-contextual, epistemic reasons for pursuit of a hypothesis in the same sort of way as for having a good reason to believe the hypothesis. Where Achinstein argues that we may neglect facts to get a non-objective sense of evidence, the claim here is that if we neglect pragmatic factors this will reveal an objective sense of evidence.

59

The claim is not that such reasons suffice for reasonable pursuit. Unlike the case of reasonable belief, pragmatic concerns may trump or be trumped by these non-pragmatic evidential reasons for pursuit. But what is important here is simply that they exist, that there is some sense of evidence playing some important role in the rationality of pursuit. This is enough to make it worth understanding what it is and how it works.

Achinstein may object that if this sort of evidence exists, it is not a new species, that is, it is not the evidence-for that we are seeking, but simply a degenerate application of his evidence-that applied to a weaker proposition than the truth of the hypothesis. In the context of pursuit, according to this line, we are still using the same notion of evidence, but here we are not looking for confirming evidence for the hypothesis itself but rather confirming evidence for statements *about* the hypothesis, for example, that the hypothesis is likely to be fruitful or possibly true.

Interestingly, this is not the line that Achinstein himself took in addressing this sort of question. In (1978), Achinstein examines a weaker use of "evidence" in cases such as being one of fifty finalists in a lottery, or playing Russian roulette. By the old account—and even less by the new version—Achinstein denies that a one in fifty chance of winning a million dollars or a one in six chance of dying from a self-inflicted wound is evidence of upcoming riches or demise. But he concedes that there does seem to be some sense in which these facts are evidence. The move Achinstein makes is to hold these facts not to be evidence that the hypothesis is true, but as evidence for the weaker claim that the hypothesis *is possible* or *may be true*.[7] In explicating this notion, he keeps only the form of the old conjunctive definition of evidence, altering the conditions by lowering the threshold value for the probabilities: "e is evidence that h may be true (or that h is possible) if (a) e is true; (b) e does not entail h; (c) the probability of h given e is not negligible; (d) the probability that there is an explanatory connection between h and e, given h and e, is not negligible" (Achinstein, 1978: 43).

This is the clue that we will take, but we expand upon the project in two ways. First, we argue that the vague notion that "*h* may be true" or "*h* is possible" is not a fine enough brush for our purposes of painting the workings of evidence in the context of pursuit. The vagueness of Achinstein's presentation hides several well-marked epistemological milestones that serve important roles in the non-contextual part of the discussion of reasonableness of pursuit, and when we bring these milestones to the fore and embed them in the language of Achinstein's new view, we see that there are important thresholds that are scientifically germane to reasonableness of belief and that are less stringent than good reason to believe.

Second, we will argue that this multiplicity of thresholds naturally yields other senses of evidence, thereby restoring the denied ambiguity within Achinstein's own framework. Carnap distinguishes between confirming evidence, supporting evidence, and comparative evidence, and allowing a number of thresholds allows for all three to be defined within Achinstein's framework. Achinstein's own view explicates confirming evidence. The passing of any or several thresholds by a hypothesis, in light of a given fact, would render that fact supporting evidence. Comparative evidence comes in comparing the thresholds surpassed by competing hypotheses given the same set of facts.

We will point to three easily recognizable, but epistemologically relevant mileposts that involve epistemic thresholds in Achinstein's sense, that is, they are steps in the usual scientific step function. In order of ascending strength, we have (1) the threshold where a hypothesis moves from "the fringe" to being a subject of legitimate scientific consideration; (2) the threshold at which a hypothesis is a "live option" or "serious contender" that needs to be taken seriously by practitioners in the field; and (3) the threshold at which there is not yet sufficient reason to believe the hypothesis, but there is good reason to believe that there will be. We find examples from the history of science chronicled by Achinstein himself that illustrate arguments given for the reaching of each of these thresholds that are less stringent than "good reason to believe."

Again, it is certainly true that the complete argument in support of the claim that a given hypothesis should be pursued will necessarily include pragmatic factors. The question of the reasonableness of pursuit of a hypothesis is most often elliptical for the question of the reasonableness of pursuit of the hypothesis in a given historical, political, or technological context. But these pragmatic conditions come into play only after the objective state of the hypothesis has been assessed.

That these milestones are objective, epistemic markers follows from the fact that they are capable of being clearly formulated in terms of probability statements to be understood in terms of Achinstein's own objective epistemic interpretation of the probability calculus. If Achinstein's own high-probability conditions are objective and purely epistemic as a result of this form of presentation, then so must similarly stated conditions with lower probability thresholds be. If these thresholds can be shown to be empirical and germane, it allows us to construct an account of supportive evidence that satisfies all of Achinstein's requirements and undermines his argument against ambiguity.

MAXWELL AND "A SUBJECT OF RATIONAL CURIOSITY"

The first threshold to consider is that beyond which an evidence report makes a hypothesis reasonable to consider pursuing at all. Since absurd but possible hypotheses generate no rational confidence, there will be a lowest threshold that is necessary to even put the hypothesis on the epistemological map. This threshold is characterized by the condition, $0 \ll p(\text{expl}[e,h]/e) \ll 1$. Such a condition is scientifically relevant because it gives us an objectively determinable lower bound on reasonable pursuit.

A historical example in which this threshold is the object of serious scientific discussion is furnished by Achinstein in *Particles and Waves*, where he discusses Maxwell's 1860 paper, "Illustrations of the Dynamical Theory of Gasses." A significant result from this first paper on the molecular kinetic theory of gasses is that "the theoretical viscosity equation implies that the coefficient of viscosity is independent of the density of the gas, and, therefore of its pressure, a result that Maxwell found intuitively very surprising" (Achinstein, 1991: 155). This fact was corroborated by experiments conducted in 1865 by Maxwell and his wife for various pressures and temperatures. "This experimental verification of a surprising theoretical consequence of his 1860 theory increased Maxwell's confidence that he was not simply building castles in the air" (Achinstein, 1991: 156).

Maxwell, quite rightly by Achinstein's lights, does not at all claim that these observations give any good reason to believe in the truth or probability of a molecular kinetic account of heat. Maxwell did, after all, also derive consequences in the 1860 paper that were inconsistent with observation, such as the proportionality of the coefficient of viscosity with the root of temperature. Such problems were not unexpected by Maxwell, since a central assumption of the derivations in 1860—that molecules are "small, hard, and perfectly elastic spheres acting on one another only during impact" (Achinstein, 1991: 155)—was believed by Maxwell to be false from the start and was only employed for ease of calculation.

But while Maxwell does not take the observations to be evidence that the kinetic hypothesis is true or likely, he does use these observations as a central part of an empirically based rational argument. We can take the inclusion of reports of these observations in an 1866 Bakerian lecture as reason to see that Maxwell was putting this empirical data forward as more than a merely subjective reason in support of the project. Maxwell makes an explicitly rational appeal for the reasonableness of pursuit, that is, a rational argument that has the rather weak aim to "lay the foundation of such investigations" (Achinstein, 1991: 155). In light of his own observations, as well as those of Daniel Bernoulli, Clausius, Joule, and others, Maxwell argues that the hypothesis that heat is a mechanical phenomenon "becomes a subject of rational curiosity" (Achinstein, 1991: 151).

Maxwell's argument can be interpreted in the current framework as asserting that on the strength of certain facts, the kinetic theory has passed an objective epistemic threshold and therefore deserves to be considered a legitimate route of scientific concern, even if none of those facts garners good reason for belief in it. There is no doubt that Maxwell's argument as a whole contains essentially pragmatic considerations. His concern is, after all, to determine whether or not the kinetic theory of gasses ought to be pursued in the 1860s. But the pragmatic concerns are built on top of the objective state of the explanatory connection, given the observed facts of the world. If the facts had been different, the likelihood of the explanatory connection would be different, having a real effect on the reasonableness of pursuit of the hypothesis even if the contextual factors are considered invariant. The facts that take the hypothesis over this threshold deserve the title "evidence" because they provide objective, empirical support to the hypothesis in the form of good reasons to consider the hypothesis a matter of legitimate scientific concern.

BOHR AND "FUTURE DEVELOPMENT OF OUR UNDERSTANDING"

The second threshold is stronger than the first, in that observation statements do not merely put the hypothesis in play epistemically; they further show that the hypothesis must be taken seriously by those in the field because it is nearly likely to be true or nearly likely to posit a legitimate explanatory connection between the hypothesis and the observation statement. We can characterize this threshold with the condition, $p(expl[h,e]/e) < 1/2$, but $p(h/e) \approx 1/2$ or $p(expl[h,e]/h,e) \approx 1/2$. Such a threshold is scientifically germane. Whereas the first threshold wins the hypothesis a spot on the team, this threshold shows that the hypothesis is off the bench and in the game. The facts that would have been evidence under Achinstein's older conjunctive definition of evidence—but which fail under the new, stronger, multiplicative definition—would be evidence here in allowing a hypothesis to cross this threshold.

In his article "How to Defend a Theory Without Testing It," Achinstein provides us with a historical example of an argument for the passing of such a threshold during his discussion of Niels Bohr's 1913 trilogy of papers, "On the Constitution of Atoms and Molecules." In these papers, Bohr derives his famous picture of the hydrogen atom through quantization assumptions, including: (1) Radiation is absorbed and emitted in discrete amounts and only when the system undergoes a state transition; (2) The laws of classical mechanics hold for the states of the atom, but not during state transitions; (3) During the transition the emitted radiation has a unique frequency satisfying Bohr's equation $E = h\nu$; and (4) If we assume circular electron orbits, the electron's angular momentum is quantized in multiples of h. From this model Bohr derives values that are in general agreement with observations for the radius of an electron's orbit, the frequency of an electron's revolution, the ratio of binding energy to the charge of an electron, and the Rydberg constant, in addition to allowing for the derivation of formulas for the Balmer and Paschen series.

Like Maxwell, Bohr took these experimental results to be an important part of an argument connected to the objective reasonableness of the hypothesis, but not as confirming evidence in Achinstein's sense. In a letter to Hevesy concerning the theory, Bohr says that the derived results give him "hope to obtain a knowledge of the structure of the systems of electrons surrounding the nuclei in atoms and molecules, and thereby a hope of a detailed understanding of what we may call 'chemical and physical' properties of matter . . . I am sure that you will understand . . . that I don't speak of the result which I mean I can obtain by help of my poor means, but only the point of view—and hope to and believe in [sic] a future (perhaps very soon) enormous and unexpected?? development of our understanding—which I have been led to by considerations as those above" (quoted in Achinstein, 1993: 98). As Achinstein puts it, "The initial explanatory success of his theory makes it reasonable to hope that knowledge of the structure of the atom will be obtained by pursuing the theory. It does not show that it has been" (Achinstein, 1993: 98).

In his three papers, Bohr mounted a vigorous defense of his atomic model based on these empirical results. But while the conclusion of this argument does not assert full confirmation of the theory, it does argue for a status stronger than Maxwell's "subject of rational curiosity." Bohr is not arguing merely that this is a path worth considering setting out upon, or that we have reached the end of the path, but that it is a median position in which we now have some empirical reason to think that this will be the right path in the end, that is, some hope to "believe in a future development of our understanding" as a result of working through this theory. We again have observation statements that are being employed to argue for the crossing of an objectively-placed epistemic threshold. As such, these empirical results deserve the designation "evidence."

Perrin and the Edge of Reasonable Belief

The third threshold is the strongest in the context of pursuit. Crossing this threshold places a hypothesis on the plateau just before good reason to believe. It is the case in which a hypothesis is awaiting one last push to be reasonable to believe. This threshold can be characterized by the condition, $p(expl[h,e]/e) < 1/2$, but $p(expl[h,e]/e) \approx 1/2$. It is the case in which the back of the epistemic camel awaits one last empirical straw.

Achinstein provides us with a historical example of this sort in his article "Jean Perrin and Molecular Reality." In this article, Achinstein (1994) sketches out in detail the argument presented by Perrin, which gives the sort of confirming evidence for the existence of molecules that satisfies Achinstein's conditions—both the older and the newer sets. Perrin's argument is grounded in the experimental work he did deriving a value for Avogadro's number from careful observation of Brownian motion employed as an analogue of an ideal gas, and then comparing that value to other experimentally determined values of Avogadro's number from a number of quite different approaches. The agreement of all such results is taken to imply that the value of Avogadro's number has been conclusively established, and since Avogadro's number is a number of molecules, there must be molecules to have a number of them.

64

Achinstein takes Perrin's observational results to satisfy the multiplicative condition for evidence. As a result, Perrin's work gives good reason to believe in the existence of molecules. But what is of interest in our context is the argumentation that Perrin gives before and after his discussion of the Brownian motion experiment. For his work on Brownian motion to be the last push that puts the molecular hypothesis on firm epistemic footing, he must argue that the state of the hypothesis before his contribution had passed our third threshold.

In the argument, Perrin compares the value he derives for Avogadro's number with those derived from Einstein's consideration of the mean free path of Brownian particles, considerations of alpha decay, observations of x-ray diffraction, considerations of blackbody radiation, and results of electrochemical phenomena. Knowing about all of these results, as well as previous work on molecular explanations of Brownian motion, Perrin contends that "the hypothesis of molecular agitation gave an admissible explanation of Brownian movement, but that no other cause of the movement could be imagined, which especially increased the significance of the hypothesis" (Perrin, 1901: 510–11). That is to say, he considers both the probability of truth of the molecular hypothesis and the probability of the explanatory connection of the molecular hypothesis with the observation of Brownian motion to be high. Yet the need was still present to include his own very exacting experiments, indicating that before these experiments were carried out Achinstein's "good reason to believe" threshold had not yet been crossed.

There is no doubt that in the subjective sense, Perrin already took this previously existing information as *his* evidence. But it was his work that eventually swayed even some of the most critical opponents of the hypothesis, such as Wilhelm Ostwald. A reasonable debate can be had about whether the likes of Ostwald and Ernst Mach opposed the atomic hypothesis out of line with reason, but this discussion presupposes the ability to talk about what earlier observations set up the observations that made it no longer rational to deny the hypothesis. Even if some subset of the results cited by Perrin did not give good reason to believe, independent of the Brownian motion results, then some proper subset of this subset would not either, although it would come close to doing so.

We may speak not only about Perrin's work having scored the epistemological goal, but also about the other observation statements that got the epistemological assist. It seems that those observation statements whose shoulders were stood upon in order to see so far play an essential part in the scientific process, and despite not providing evidence in and of themselves that the hypothesis is true, they deserve to be called "evidence."

Restoring Ambiguity

Achinstein's own case studies, therefore, provide us with three examples of objectively definable thresholds—less stringent than good reason to believe—that have played key roles in important episodes in the history of science. We can use this multiplicity of epistemic thresholds to define supporting evidence for *h* as any fact *e* such that given *e*, *h* is taken over at least one epistemological threshold.

The claim here is not that every scientific hypothesis must proceed across each of these thresholds one at a time. There are certainly examples of hypotheses for which a single fact is sufficient to surmount all mentioned thresholds, providing good reason to believe. Finding fresh bear dung on a trail provides immediate confirming evidence of bears in the area. But this is not always the case in science. There are many cases for which the passing of one or another of these earlier thresholds is a matter of legitimate debate within the scientific community, and the basis often given for the claim of having passed the threshold in question is the results of empirical observation independent of the pragmatic concerns otherwise operative in the context of pursuit.

It might be objected that the threshold concept itself is inherently inconsistent with the very notion of supporting evidence, that crossing a threshold is a discrete notion not compatible with the seemingly smooth concept of increased support. On this line, the existence of weaker thresholds only posits additional forms of confirming evidence. We may now have a variety of levels of evidence-that for propositions involving the hypothesis such as "h is likely" or "h is close to being confirmed," but this is not what we mean by "supporting evidence."

This objection may be answered in two ways. First, the claim is not being advanced that the three thresholds that have been discussed are a complete list of thresholds, or even the most important in the context of pursuit. There may in fact be many more gradations when scientific argumentation is carefully examined. As Achinstein (2001: 83) points out, as additional steps are added to the step function, we see the classical view of supportive evidence appear as an approximation. The relation between Carnap's classical version and the new quantized account is something of an epistemological version of the correspondence principle.

66

Second, there is a very crucial difference between the notion of evidence-that, for lower thresholds, and evidence-for. It is an important feature of the threshold picture that once one has passed a threshold, facts that "make progress along the plateau" are still considered evidence-that with regard to the passing of that threshold. Once there is good reason to believe a hypothesis, other observations may use that likelihood as a foundation, and one can more easily accrue further good reasons for belief after the "epistemological glass ceiling" has been broken. In the same way, if we consider a hypothesis for which we have evidence-that it has passed, say, our first threshold and we acquire a new fact that also speaks to its being Maxwell's "subject of rational curiosity" but is insufficient to propel the hypothetical hypothesis over our second threshold, then that proposition is evidence-that for the first threshold—but it is *not* evidence-for the hypothesis. Although evidence-for a hypothesis will necessarily be evidence-that it has crossed some particular threshold, the converse does not hold. Simply because a fact is evidence-that does not necessarily make it evidence-for the hypothesis. Hence, there is a real distinction to be drawn between evidence-for and evidence-that. Supporting evidence cannot therefore be subsumed under the notion of confirming evidence by considering conclusive evidence for weaker propositions. Rather, it is a different species altogether.

Consequently, we can reasonably speak of increasing support for the hypothesis despite (1) the discontinuous picture of reasonable belief from Achinstein that is being presupposed, and (2) the fact that we may not have any good reason to believe the hypothesis. To give the name "evidence" to the objective, empirical, scientifically relevant relation between such hypotheses and facts not only is reasonable, but—contrary to what he has argued—is perfectly in line with all of Achinstein's constraints on a successful philosophical account of scientific evidence.

ACKNOWLEDGMENT
I wish to sincerely thank Peter Achinstein for his patience in conversation and his occasionally legible, but always invaluable comments on earlier drafts of this chapter.

NOTES
1. Achinstein rejects the standard notion of degree of belief in a proposition and replaces it with the degree of reasonableness of belief in the hypothesis. See (2001), pp. 118–20.

2. Achinstein insists that evidence requires a relation between facts and hypotheses, and not observation reports and hypotheses.

3. The argument on pp. 114–24 of (2001) is a slightly revised version of that given in (1983b), pp. 328–34.

4. An explanatory connection exists if h correctly explains why e is true, e correctly explains why h is true, or there exists a third statement that explains why both are true. Let expl[h,e] represent the proposition "There exists an explanatory connection between h and e."

5. Two additional conditions, that the evidence statement is true and that the evidence statement does not entail the hypothesis, were also posited as necessary.

6. The probability statements in Achinstein's definition are conditional, assuming the truth of the hypothesis and evidence statement.

7. This notion should not be confused with Achinstein's notion of potential evidence, in which the conditions for evidence are satisfied, but the truth of the hypothesis and existence of the explanatory connection are not guaranteed.

REFERENCES
Achinstein, P. (2001) *The Book of Evidence*. New York: Oxford University Press.
Achinstein, P. (1994) "Jean Perrin and Molecular Reality." *Perspectives on Science* 2, 396–427.
Achinstein, P. (1993) "How to Defend a Theory Without Testing It: Niels Bohr and the 'Logic of Pursuit'." *Midwest Studies in Philosophy* XVIII, 90–120.
Achinstein, P. (1991) *Particles and Waves*. New York: Oxford University Press.
Achinstein, P. (1983a) *The Nature of Explanation*. New York: Oxford University Press.
Achinstein, P. (1983b) *The Concept of Evidence*. New York: Oxford University Press.
Achinstein, P. (1978) "Concepts of Evidence." *Mind* 87, 22–45.
Carnap, R. (1962) *Logical Foundations of Probability*. Chicago: University of Chicago Press.
Kaplan, M. (1996) *Decision Theory as Philosophy*. Cambridge: Cambridge University Press.
Perrin, J. (1901) "Brownian Movement and Molecular Reality." In *The Question of the Atom*. M. Nye, ed., 1986. Los Angeles: Tomash Publishers, pp. 507–601.
Salmon, W. (1975) "Confirmation and Relevance." In *Induction, Probability, and Confirmation*, G. Maxwell and M. Anderson, eds. *Minnesota Studies in the Philosophy of Science* 6. Minneapolis: University of Minnesota Press, pp. 5–36.

THE FAST TRACK TO CONFIRMATION:
ACHINSTEIN AND PEIRCE ON EVIDENCE

Frederick M. Kronz and Amy L. McLaughlin

Peter Achinstein's *The Book of Evidence* (2001) begins by recounting a challenge made to him by the dean of his college; namely, to come up with a theory of evidence that will be of use to working scientists. This is a legitimate challenge. Philosophers are notorious for developing theories that are only of interest to other philosophers (if anyone). We believe that Achinstein has risen to the challenge in developing his new theory of evidence. His analysis effectively captures in a rigorous manner what working scientists typically mean by "evidence." He also explains convincingly why other attempts to develop a viable theory of evidence fall short of this goal. Consequently, his theory will be of interest both to working scientists and to philosophers.

The issue that we address here is how to make his theory of evidence more workable for scientists. It will suffice, for the purpose at hand, to give a brief characterization of his notion of evidence. The pre-philosophical notion is this: Evidence for a hypothesis is that which provides a good reason for believing the hypothesis. Alternatively, it is that which may be said to confirm, corroborate, substantiate, authenticate, validate, or verify the hypothesis. Achinstein's explication of this notion requires first that evidence for a hypothesis must make it probable that the hypothesis is true, meaning that the probability of the hypothesis—given the evidence—must be greater than one-half. He then demonstrates that this is a necessary, but not a sufficient condition for evidence. There must also be a high probability that there is an explanatory connection between the evidence and the hypothesis (given the evidence and the hypothesis). Moreover, the product of these two probabilities (of the hypothesis given the evidence, and of there being an explanatory connection between them) must be greater than one-half. This clearly is a very robust and demanding notion of evidence.

For what follows, evidence may be regarded as a substantial body of data that satisfies the conditions specified by Achinstein. Of course, there may in general be many different bodies of data that satisfy these conditions with respect to the hypothesis at hand, and thereby serve as evidence for the hypothesis. This fact gives rise to the following question: How should a scientist go about collecting data in order to provide evidence for a hypothesis? Achinstein's theory does not address this question. It specifies when a body of data counts as evidence, but it does not indicate how to go about putting together such a body. That is to say, Achinstein's definitional theory needs to be supplemented by a pragmatic codicil.

70

The primary reason that this supplement is desirable is the challenge to develop a theory of evidence that can be used to good effect by working scientists. Knowing when and why there is evidence for a hypothesis is certainly useful, but even more useful is knowing the most efficient ways to go about obtaining a body of data that will constitute evidence for hypotheses when such evidence is lacking. Efficiency is an important matter for the working scientist, since key resources are limited. It is a pragmatic matter that falls under the rubric of the economy of research. Peirce, the founder of pragmatism, was very concerned about this and related issues: "Research must contrive to do business at a profit; by which I mean that it must produce more effective scientific energy than it expends. No doubt, it already does so. But it would do well to become conscious of its economical position and contrive ways of living upon it" (CP 7.159).[1]

The limited resources that he repeatedly mentions in emphasizing the need for efficiency are time, money, energy, and thought. He had some important insights about how to conduct scientific inquiry so that resources are used most efficiently. In the sections that follow, we elaborate key aspects of Peirce's theory of scientific inquiry and the economy of research.[2] These aspects of Peirce's system are explicated and then put forward as a friendly amendment (by addition) to Achinstein's theory. The Peircean account does not require the acceptance of Achinstein's theory; it could be used in conjunction with any theory of evidence that does not undermine Peirce's commitments. At present, however, Achinstein's theory of evidence is the most compelling of the alternatives, especially in light of his criticisms of the standard theories.

THE ECONOMY OF RESEARCH

Charles Peirce is best known for his contributions to philosophy—notably his pragmatism and semiotics. But his primary training and work experience were in science. His shift to philosophy was not a result of failed scientific prowess; on the contrary, he was quite successful in his scientific endeavors. Among his major scientific advancements are the following: reformation of stellar magnitude scales with the aid of instrumental photometry; discovery of an error in European gravitational measurements due to flexure of the pendulum stand, and the subsequent invention of a superior pendulum; determination of the length of the standard meter in terms of light wavelengths; development of a new form of map projection (a quincuncial map) using elliptical functions; and collaboration in the development of hydroelectric power. But philosophy was Peirce's first love. Indeed it was his interests in logic and the scientific method that compelled him to begin his scientific investigations. When he was focused on his philosophical work, that work was fundamentally informed by his scientific expertise.

According to Peirce, scientific inquiry begins when one encounters a challenge to one's existing set of beliefs. A challenge can present itself when this set entails that some existent phenomenon either (1) should not have occurred, (2) should have occurred differently than it has, or (3) cannot be sufficiently accounted for.[3] This tension between theory and observation demands a resolution, and a first step toward providing one is to develop a hypothesis that would (if true) serve to explain the surprising phenomenon as a matter of course. It is not unusual for scientists to consider several explanatory hypotheses at first, and then use a process of elimination that eventually focuses research on just one of the alternatives.

Peirce uses the term *abduction* to denote an inference from some surprising phenomenon to an explanatory hypothesis.[4] It is one of three modes of inference that he regards as crucial to scientific inquiry. The other two modes are deduction and induction, and they are used in concert to promote the explanatory hypothesis (obtained in abduction) from being merely conjectural to being well-confirmed. Usually, more than one explanatory hypothesis is under consideration, but this complication will be ignored for the moment. According to Peirce, the process typically proceeds as follows. Observational consequences are deduced from a hypothesis (together with some well-established auxiliary assumptions). A suitable selection of these is made and then tested empirically. So long as the tests are positive, the hypothesis is retained. If some are negative, then the hypothesis may be rejected, in which case the process of inquiry continues with an abductive inference to a new explanatory hypothesis, which may range from being a suitably modified version of the rejected hypothesis to being a radically different hypothesis.

The account of scientific inquiry as outlined above has some well-known difficulties that some, perhaps many, regard as being insurmountable. In particular, the account invokes two complementary scientific methodologies, hypothetico-deductivism and falsificationism,[5] that are regarded by some as dinosaurs of the departed positivist age.[6] What we are concerned with here, though, are Peirce's recommendations for testing hypotheses. Peirce's descriptions invoke aspects of his methodological principles but do not depend crucially upon them, so it is reasonable to proceed with the analysis under the Peircean scenario (even if one regards Peirce's methodological principles as notorious rather than notable).

72

Suppose that scientists have put forth several hypotheses to explain some surprising phenomena. The key question is how to go about obtaining data that will rule out some of the hypotheses and support others. A common response is that one should derive observational consequences from each hypothesis and then test them experimentally, but this response must be supplemented with more practical considerations. There may be an abundance of empirical consequences that one might derive from a hypothesis. Where there are several hypotheses under consideration, it is even more difficult to choose which effects are to be explored experimentally.

The key problem is to formulate criteria to guide scientists in making judicious choices for experimentally testing hypotheses. Peirce solves this problem by appealing primarily to economic considerations. That is to say, he appeals to the bottom line: It is necessary to make judicious choices based primarily on considerations of resource allocation. He states, "[I]n view of the fact that the true hypothesis is only one out of innumerable possible false ones, in view, too, of the enormous expensiveness of experimentation in money, time, energy, and thought, (a necessary consideration) is the consideration of economy. Now economy, in general, depends upon three kinds of factors: cost; the value of the thing proposed, in itself; and its effect upon other projects" (CP 7.220).

The goal, as Peirce sees it, is to narrow the field of viable hypotheses by making efficient use of the limited available resources. It is to be accomplished by carrying out a cost/benefit analysis. Peirce elaborates on this theme as follows: "The doctrine of economy, in general, treats of the relations between utility and cost. That branch of it which relates to research considers the relations between the utility and the cost of diminishing the probable error of our knowledge. Its main problem is, how, with a given expenditure of money, time, and energy, to obtain the most valuable addition to our knowledge"[7] (CP 7.140). In this passage, Peirce subsumes epistemic considerations under the rubric of economy. Epistemic value will factor prominently, though certainly not exclusively, in the cost/benefit analysis.

The view expressed in the foregoing passages concerning the role of economic factors in conducting research is programmatic at best. Its application requires the introduction of specific guidelines that serve to develop the main idea. Though Peirce offers several guidelines of this sort, we focus on those that are most central to the main issues of this chapter. One of Peirce's recommendations concerns how to facilitate early elimination of some of the alternative hypotheses: "[I]f a hypothesis can be put to the test of experiment with very little expense of any kind, that should be regarded as a recommendation for giving it precedence in the inductive procedure. For even if it be barely admissible for other reasons, still it may clear the ground to have disposed of it" (CP 7.220).

73

Giving precedence in the inductive procedure to one hypothesis among several explanatory hypotheses means putting that hypothesis to the test first. Peirce recommends giving precedence to that explanatory hypothesis for which experimental testing involves very little expense (though he gives no explicit lower-limit cost threshold).[8] It is worth emphasizing that Peirce's recommendation is at variance with what might otherwise be expected: that one should test hypotheses that appear most likely to be true. He elaborates on this theme elsewhere, as follows: "In the light of one's metaphysics and general conception of the department of truth dealt with, one considers what different hypotheses have any claims to investigation. The leading considerations here will be those of the 'economics' of research. If, for example, a hypothesis would necessitate an experimental result that can be cheaply refuted if it is not true, or would be greatly at variance with preconceived ideas, that hypothesis has a strong claim to early examination" (CP 7.83).

Peirce recommends giving precedence to frugal experiments for two reasons: to possibly rule out some of the alternative hypotheses, and to possibly support those that are greatly at variance with preconceived ideas. These two types of cases confer the most valuable additions to our knowledge that may be obtained from such experiments. If a viable hypothesis can be cheaply refuted, this confers an obvious advantage by narrowing the field of investigation, thereby more efficiently focusing resources. If a hypothesis is suitably explanatory but counterintuitive, it is worth devoting minimal resources to explore its viability, since a slim measure of support provides considerable potential for conceptual revolution. Presumably, giving support to what is regarded as a likely hypothesis or ruling out an unlikely hypothesis is valuable, but to a much lesser degree.

In the following passage, Peirce indicates that the apparent likelihood of a hypothesis is the predominant reason given for experimentally testing it.

Of course, if we know any positive facts which render a given hypothesis objectively probable, they recommend it for inductive testing. When this is not the case, but the hypothesis seems to us likely, or unlikely, this likelihood is an indication that the hypothesis accords or discords with our preconceived ideas; and since those ideas are presumably based upon some experience, it follows that, other things being equal, there will be, in the long run, some economy in giving the hypothesis a place in the order of precedence in accordance with this indication. (CP 7.220)

Peirce is making the point that "other things being equal," one may appeal to apparent likelihood in prioritizing hypotheses for experimental testing. The "other things" he alludes to are other economic factors, and these turn out to be the predominant ones for Peirce. As he continues, he expresses serious reservations about the significance of apparent likelihood, and gives preference to other economic factors (which are discussed later):

But experience must be our chart in economical navigation; and experience shows that likelihoods are treacherous guides. Nothing has caused so much waste of time and means, in all sorts of researches, as inquirers' becoming so wedded to certain likelihoods as to forget all the other factors of the economy of research; so that, unless it be very solidly grounded, likelihood is far better disregarded, or nearly so; and even when it seems solidly grounded, it should be proceeded upon with a cautious tread, with an eye to other considerations, and a recollection of the disasters it has caused. (CP 7.220)

The upshot is that Peirce regards considerations of the economy of research other than apparent likelihood (such as the cost and the potential benefit of the experimental test) as deserving greater weight in determining which hypothesis should be tested first.[9] Moreover, once the testing of a hypothesis has proven favorable to it, that outcome can then serve to provide substantial incentive (according to Peirce) for further testing of the hypothesis. This means, in part, that assessments of likelihood become trustworthy to the extent that successful experimental testing has solidly grounded it.[10]

Though one may be inclined to think of the strength of a hypothesis in terms of its likelihood, Peirce measures the strength of a hypothesis in terms of the resources that the scientific community is willing to devote to experimentally testing that hypothesis.[11] He refers to this in more neutral terms as the urgency of the hypothesis: "Perhaps we might conceive the strength, or urgency, of a hypothesis as measured by the amount of wealth, in time, thought, money, etc., that we ought to have at our disposal before it would be worth while to take up that hypothesis for examination. . . . The expense which the examination of it would involve must be one of the main factors of its urgency" (CP 2.780). The strength (or urgency) of a hypothesis will grow as it continues to be tested and as its competitors are eliminated by way of experimental testing. As the process continues, one hypothesis will rise to the fore. Its urgency grows and the scientific inquiry continues until confirmation is achieved. Until then, it is to be expected that Peirce would require considerations of economy to continue as a guide to scientific inquiry. This expectation is vindicated in the next section.

The Fast Track to Confirmation

75

Suppose that scientists put forth several hypotheses to explain some surprising phenomena. All but one of the hypotheses have been ruled out experimentally; nevertheless, the body of data gathered thus far fails to serve as evidence for the remaining hypothesis. According to Peirce, this situation is not particularly unusual. Scientists are often in possession of data and a hypothesis that explains the data without the data's being evidence that the hypothesis is true (or approximately true).[12] The key question is how to go about obtaining additional data to suitably complement the initial data, so that the combination constitutes evidence for the hypothesis. That is to say, what criteria will guide scientists to judicious choices as to which experimental tests to perform? Peirce suggests an effective answer to this question in the following passage.

> Experiment is very expensive business, in money, in time, and in thought; so that it will be a saving of expense, to begin with that positive prediction from the hypothesis which seems *least likely to be verified*. For a single experiment may absolutely refute the most valuable of hypotheses, while a hypothesis must be a trifling one indeed if a single experiment could establish it. When, however, we find that prediction after prediction, notwithstanding a preference for putting the most unlikely ones to the test, is verified by experiment, whether without modification or with a merely quantitative modification, we begin to accord to the hypothesis a standing among scientific results. (CP 7.206, *emphasis added*)

Peirce indicates that considerations of economy compel the scientist to pay particular attention to the predictions of the hypothesis that are least likely to be verified. Such predictions should be among the first to be considered for experimental testing to ascertain whether they are true. Of course, once such predictions are obtained, other considerations of economy should come into play.[13] For example, it may not be clear at all how to go about testing the least likely to be verified prediction, and much expense in time and thought may be required to determine how to do so. Moreover, even if one were able to conceive of a suitable test, the test might be prohibitively expensive (again, broadly construed). So, it may be decided that the prediction that is least likely to be verified is not to be tested, in which case one considers next the prediction that is least likely to be verified of those remaining.

Peirce's phrase "least likely to be verified" is in need of further analysis. The key question concerns the information upon which the probabilistic assessment is to be made. A related passage sheds some light on this matter: "The verification . . . must consist in basing upon the hypothesis predictions as to the results of experiments, especially those of such predictions *as appear to be otherwise least likely to be true,* and in instituting experiments in order to ascertain whether they will be true or not" (CP 7.89, *emphasis added*). This passage contains a term that serves to focus the question now under consideration. Peirce indicates that attention should be focused especially on those predictions that would "appear to be otherwise least likely to be true," the key term being "otherwise." There are at least two different ways to understand this qualification. First, it could mean upon regarding the hypothesis as false. Second, it could mean upon omitting consideration of (i.e., upon conceptually setting aside) the hypothesis in question.

Support for both readings is present in Peirce's writings, and the following passage clearly supports the first one: "A hypothesis having been adopted on probation, the process of testing it will consist, not in examining the facts, in order to see how well they accord with the hypothesis, but on the contrary in examining such of the probable consequences of the hypothesis as would be capable of direct verification, especially those consequences which would be very unlikely or surprising *in case the hypothesis were not true*" (CP 7.231, *emphasis added*).

In contrast is a passage that directly supports the second reading, particularly in the last sentence.

76

> [One] line which our studies of the relation of the hypothesis to experience may pursue, consists in directing our attention . . . [to] what effect that hypothesis, if embraced, must have in modifying our expectations in regard to future experience. Thereupon we make experiments . . . in order to find out how far these new conditional expectations are going to be fulfilled. In so far as they greatly modify our former expectations of experience and in so far as we find them, nevertheless, to be fulfilled, we accord to the hypothesis a due weight in determining all our future conduct and thought. The strength of any argument (from the fulfillment of a prediction to a hypothesis) depends upon how much the confirmation of the prediction runs counter to what our expectation would have been *without the hypothesis.* (CP 7.115, *emphasis added*)

Although the two interpretations of "otherwise" appear to be quite different, there is a third interpretation that is consistent with each. It derives from a particular way of understanding Peirce's pragmatic maxim. Peirce offers several formulations of the pragmatic maxim, and one of the mature forms he develops is described as follows: "[T]he maxim of pragmatism is that a conception can have no logical effect or import differing from that of a second conception except so far as, taken in connection with other conceptions and intentions, it might conceivably modify our practical conduct differently from that second conception" (CP 5.196).

A key point revealed in this passage is Peirce's recognition that a comparison among alternatives is a (perhaps *the*) vehicle for elucidating empirical consequences. He states that the pragmatic maxim is intended to serve as a criterion for judging the significance of a hypothesis *as compared with that of another hypothesis*. Thus, we construe Peirce's condition as follows: When one seeks to verify a hypothesis, h, one should seek out those experimental effects of h that appear least likely to be true *on the assumption of some competing, contrary hypothesis*—the one that would be superseded if h were confirmed, which we denote herein as h'.[14] This interpretation serves to reconcile the other two as follows. Since h' and h are contraries, h' entails ~h; so, assuming h' is tantamount to setting aside h as well as assuming ~h.

This third interpretation, invoking some competing (contrary) hypothesis, has another advantage. It allows one to talk about the likelihood of empirical predictions. By contrast, this is not possible with either of the other interpretations. In the case where h is merely set aside without some alternative hypothesis to serve as a guide, one cannot even formulate predictions, let alone assess their likelihood. In the case where h is considered false *simpliciter* (i.e., without any further specification), the resulting hypothesis would not serve as a suitable guide for experimentation.[15]

In what follows, we refer to the Peircean recommendation for hypothesis testing and confirmation as the "Peircean test condition." To facilitate expressing this condition formally, we use the following notation:

h   the hypothesis under consideration

b   the relevant background information

h'   a subset of b that constitutes a hypothesis that is contrary to h, and which was held just prior to the formulation of h

E   the set $\{e_1, \ldots, e_N\}$ of observational consequences non-trivially deduced[16] from the conjunction of h with b that cannot be derived from b alone[17]

E is here regarded as some finite set whose elements are under consideration for experimental testing by a scientist or a community of scientists over a period of time, meaning that we allow for the possibility that E be supplemented or revised as experimental testing progresses. We can now state the condition as follows:

> *Peircean Test Condition:* Choose the least likely $e_k \in E$ given b (the $e_k \in E$ such that $P(e_k | b) = \text{Min} \{P(e_j | b) \mid 1 \leq j \leq N\}$).[18] Assess the economy of testing for e_k (in terms of resources, etc.). If testing for e_k is cost-effective, then perform the experiment; otherwise, choose the least likely element of $E - \{e_k\}$ and proceed as before. This procedure is to be repeated on the remaining empirical consequences until the hypothesis is confirmed or disconfirmed.[19]

We regard this characterization as both programmatic and provisional. It is programmatic since the notion of cost-effectiveness requires a suitable analysis. Among other things, such an analysis will include guidelines for calculating a cost/benefit ratio and appropriate thresholds—maximum cost, maximum cost/benefit ratio, appropriate margin of error for probability calculation, and so forth. One reason that this account is provisional is that there are difficult cases that might require the condition be refined to accommodate them. A possible case of this sort is considered below.

It is possible that there are two distinct elements of E, e_1 and e_2, such that e_1 is by far the least likely element of E and is moderately cost-effective, but e_2 is substantially more cost-effective than e_1. On the Peircean model outlined here, it would seem that we must prefer to test for e_1, thus neglecting a substantially more cost-effective alternative.[20]

Peirce would likely respond to this objection by explaining that giving preference to the least likely consequence of a hypothesis already takes into account some considerations about cost-effectiveness. Peirce asks that we give preference to the least likely consequences precisely because of the potential benefit that could accrue from their consideration. If an unlikely consequence were found to obtain, we would be likely to find implications from this for other projects. Since b, the relevant background information admitted prior to the formulation of h, will set the parameters for determining the least likely consequence, and since b includes other scientific hypotheses and projects, the possible rippling effect throughout b of such a discovery may be quite dramatic. The potential benefit derived from such a rippling effect is thus built into the process. Structuring the process in this way dramatically reduces, on average, the time and energy required to calculate cost-effectiveness.[21]

Another reason we take the foregoing characterization of the Peircean Test Condition to be provisional is the complex nature of actual scientific inquiry. Scientists often are in consideration of more than just a single explanatory hypothesis, though it is rare that more than a few hypotheses are in contention at any given time.[22] In what follows, we consider the next simplest case—where two explanatory hypotheses remain under serious consideration. Peirce does not explicitly address this sort of situation. The generalization that follows is a natural extension of the account developed above.

To facilitate expressing the more general condition formally, we use the following notation:

g the first of two hypotheses under consideration

h the second of two hypotheses under consideration

b the relevant background information

h′ a subset of b constituting a hypothesis that is contrary to both g and h, and which was held just prior to the formulation of g and h

D the set $\{d_1, \ldots, d_M\}$ of observational consequences non-trivially deduced from the conjunction of b with g that cannot be derived from b alone

E the set $\{e_1, \ldots, e_N\}$ of observational consequences non-trivially deduced from the conjunction of b with h that cannot be derived from b alone

Similar to the previous formulation, D and E are here regarded as finite sets whose elements are under consideration for experimental testing by a scientist or a community of scientists over a period of time, meaning that we allow for the possibility that D or E will be supplemented or revised as experimental testing progresses. We can now state the condition as follows:

> *Generalized Peircean Test Condition:* Choose the least likely $k \in D \cup E$ given b, the k such that $P(k \mid b) = \text{Min}\{P(k \mid b) \mid k \in D \cup E\}$. Assess the economy of testing for k (in terms of resources, etc.). If testing for k is cost-effective, then perform the experiment; otherwise, choose the least likely element of $D \cup E - \{k\}$ and proceed as before. This procedure is to be repeated on the remaining empirical consequences until the hypothesis is confirmed or disconfirmed.

It should be reasonably clear how to generalize this condition further in order to accommodate the consideration of n hypotheses, for n > 2.

RELATION TO THE EXPERIMENTAL TURN IN PHILOSOPHY OF SCIENCE

The developments introduced above mesh rather well with a recent movement in the history and philosophy of science. In the early 1980s a sizable number of philosophers and historians of science shifted focus from theory to experiment. Among them are Robert Ackermann, Nancy Cartwright, Allan Franklin, Peter Galison, Ronald Giere, and Ian Hacking. The resulting literature on experiment consists mostly of interesting and informative case studies. Deborah Mayo characterizes it this way: "Their experimental narratives offer a rich source from which to extricate how reliable data are obtained and used to learn about experimental processes. Still, nothing like a systematic program has been laid out by which to accomplish this. The task requires getting at the structure of experimental activities and at the epistemological rationale for inferences based on such activities." (Mayo, 1996)

80

After criticizing the New Experimentalists and the Bayesians for reverting to theory-dominated philosophies, Mayo proposes a philosophy of experiment based on learning from error. Her approach involves the use of the error statistics methods of Fisher, Neyman, and Pearson unified into a philosophy of experiment that is strongly influenced by Peirce. She briefly discusses some of Peirce's remarks on economy of research, but does little to promote this aspect of his views.

There is one New Experimentalist who has promoted the idea of economy of research, and that is Allan Franklin. He was trained as an experimentalist in high-energy physics, but in the 1970s shifted focus to the history and philosophy of science, particularly the role of experiment in physics. He is an avowed Bayesian, and has published several papers with Colin Howson, a major advocate of Bayesian methodology. Franklin has illustrated repeatedly via case studies two key themes that fall under the rubric of the economy of research. They are:

> *Instrumental Loyalty:* Experimenters try their best to use existing apparatuses (with minor modifications, perhaps) to perform experiments

> *Recycling of Expertise:* Experimenters have substantial knowledge about how existing apparatuses work and what the experimental difficulties may be

Using existing apparatuses involves the use of existing knowledge and expertise, and this combination is most economical in terms of time, money, energy, and thought. These are certainly important insights that are worthy of incorporation into a systematic philosophy of experiment.

The experimental turn in philosophy of science seems to derive from the sorts of Peircean insights explored in the previous section. There is no doubt that a more systematic approach could be fruitfully developed from a more extensive elaboration of Peirce's considerations about the economy of research, but these are matters for future exploration. The developments discussed in this chapter lay the groundwork and constitute a first step in that direction.

CONCLUSION

Peirce provides guidelines for determining which experimental tests are to be given priority in two distinct contexts—one broad and the other narrow.[23] In both contexts, Peirce holds that considerations of economy should guide scientists to conduct those experiments that will provide the greatest value for the least expense. The broader context occurs in the initial stages of a scientific inquiry, where several explanatory hypotheses are in contention. In this context, Peirce's recommendations facilitate limiting the field of investigation. The narrower context occurs where the field has been limited to such an extent that only one hypothesis remains under consideration. In this context, Peirce provides recommendations for gathering sufficient data to confirm or disconfirm the hypothesis. In this way, Peirce offers practical advice that serves to facilitate meeting Achinstein's conditions for confirmation.

Peirce's pragmatic considerations do much to direct and focus scientific inquiry from beginning to more advanced stages. He offers little information, however, about how to determine when scientific inquiry has achieved the goal of finding a suitable hypothesis. Achinstein's account serves, in this way, to complement the Peircean approach. His account of evidence makes explicit what conditions must be fulfilled in order for a hypothesis to be considered suitably confirmed.

What we have shown, then, is that the expressed views of Peirce on the economy of research and Achinstein on the nature of confirmation serve to mutually complement one another. In order for a theory of evidence to be of use to working scientists, it must accommodate three phases of inquiry—initiation, progress toward hypothesis confirmation, and confirmation. Peirce provides the means for initiating the inquiry and reasonable guidelines for progressing toward confirmation. Achinstein provides an account of how to determine when a hypothesis is confirmed.

ACKNOWLEDGMENT

We would like to thank Peter Achinstein for commenting extensively on several incarnations of this chapter. His criticisms and suggestions have proven to be invaluable. We would also like to thank participants at the Conference on Scientific Evidence—particularly Deborah Mayo, Sharon Kingsland, and Kent Staley—for their helpful suggestions. Finally, we would like to thank Patrick Maher, Theo Kuipers, and Wayne Myrvold for their input at the Twelfth International Congress on Logic, Methodology, and Philosophy of Science.

NOTES

1. "CP" refers here and in what follows to *The Collected Papers of Charles Sanders Peirce* (1931–1935, 1958).

2. What we outline is a Peircean theory of scientific inquiry and the economy of research. We do not purport to precisely capture Peirce's own views, which were subject

to various shifts over time. Rather, we offer an interpretation of key aspects of his system that has textual support and is consistent with the primary aspects of his pragmatism.

3. Peirce typically speaks of such challenges as "recalcitrant" or "surprising" phenomena. His most explicit characterization of such a surprise occurs at CP 5.57. He indicates that someone experiencing the surprise feels a duality forced upon her. That is to say, she is confronted with a dichotomy—a choice between two contradictory alternatives. She can either accept the existence of the phenomenon, or adhere to her previous expectation set. The expectation is initially thought to be a feature of the external world, but the contrary perception leads one to recognize that the expectation is better attributed to a "mere inner world," meaning (presumably) one's own theories about the external world.

4. One may object to calling hypothesis formation an inference, preferring to regard it as an imaginative leap that is subject to psychological, rather than logical, analysis. Of course, making deductive inferences can involve an imaginative leap, but that leap can be rationally reconstructed (by providing a proof, for example). Similarly, it may be possible to give a rational reconstruction of the explanatory notion used in forming a hypothesis. Nothing in the analysis that follows depends on the possibility of providing such a reconstruction, but our instincts follow Peirce on this issue. See Lipton (1993), especially pp. 120–122, for an informative attempt to address the matter of treating this leap as an inference.

5. Here we do not regard the thesis that hypothesis formation is non-inferential as being essential to hypothetico-deductivism, though many of its proponents advocate this thesis.

6. Numerous writers discuss the shortcomings of hypothetico-deductivism and falsificationism—see, for example, Lakatos (1978), Glymour (1980), Newton-Smith (1981), Achinstein (1991), Earman (1992), or Gemes (1998). Some have jettisoned these methodologies in light of these criticisms, while others see something worth salvaging and attempt to work with them. Lakatos defends a modified version of falsificationism, and Achinstein and Gemes are (in some sense) defenders of a modified version of hypothetico-deductivism. Glymour and Earman think that hypothetico-deductivism is hopeless. The upshot is that the status of these theories is controversial.

7. We construe Peirce's characterization of cost in terms of money, time, and energy as a partial characterization. Other considerations for figuring costs (and benefits) may include social, political, professional, and personal factors.

8. Peirce suggests that if the expense of testing an explanatory hypothesis is below a certain level (what we call a lower-limit cost threshold), then the test should be performed, and further, that this suffices for recommendation, no matter how small the degree of benefit that is likely to be conferred. Some benefit will result from its being tested however the test comes out, since it would (if true) explain the surprising phenomenon. Any test whose cost is at or above the threshold level requires a cost-benefit analysis. Of course, this cost threshold is a somewhat fuzzy delineation—one might think of it in the same sort of way as perceptual thresholds, which differ from person to person and from context to context.

9. The apparent likelihood of a hypothesis appears to be closely related to what Bayesians typically refer to as the prior probability of the hypothesis. If this assessment is correct, then Peirce's apprehension with regard to apparent likelihood distances him from the Bayesian camp.

10. What Peirce means by a hypothesis being objectively probable in light of positive facts appears to be closely related to what Bayesians mean by the posterior probability of the hypothesis, given those facts.

11. Reference to the scientific community in this and related contexts is of paramount importance for Peirce; that is to say, Peirce certainly emphasizes the communal nature of scientific inquiry.

12. Peirce makes this point in distinguishing abduction from induction (e.g., see CP 8.209, CP 7.114, CP 2.96, and CP 5.171).

13. As noted earlier, Peirce regards epistemic value as an economic factor. This is one reason for alluding here to *other* economic factors, meaning those aside from epistemic value.

14. In scientific inquiry, a hypothesis is formulated to accommodate some challenge that presents itself to one's set of beliefs. It may happen that such a challenge does not speak directly to any explicitly formulated scientific theory, hypothesis, or model. Presumably then, the challenge speaks to some unarticulated set of presuppositions. In that case, h' corresponds to that set.

15. If, for example, one asserts the negation of the law of universal gravitation without further specification, the resulting statement asserts merely that there are two bodies that do not have an inverse square force between them.

83

16. A consequence of h & b is trivial, if there are elements of h and b that are mutually contradictory and the consequence essentially depends on this. We do not require that a consequence fail to be trivially deducible, only that it be deducible non-trivially. So, for example, if e_i can be trivially deduced from h & b it will still be allowed into the set E, *as long as it can be deduced from h & b non-trivially*. Also, the background information b should include any well-established theories (if any) that are associated with E, whether or not those theories are compatible with h. For this reason, we have deliberately chosen not to say that E is deducible from the conjunction of h and b-minus-h' (and not deducible from b (including h')). That is to say, excluding h' from b may not be sufficient to rule out consequences of h & b that are trivial—there could be other components of b aside from h' that are incompatible with h.

17. If all of the non-trivial observational consequences of h & b are derivable from b alone, then h fails Peirce's pragmatic maxim.

18. The considerations above entail that $P(e_k | b) < 1$ for each k (since b does not entail e_k). But it is possible for E to have no minimally probable element. In that case, one finds the least costly e_k and tests it. If these elements are equally cost-effective, then flipping a coin will do.

19. At the point of confirmation or disconfirmation, the urgency of the hypothesis subsides.

20. We could like to thank Peter Achinstein for bringing this provocative objection to our attention in private correspondence.

21. One might develop a variant of the objection raised in the previous paragraph, by supposing that there is very little difference in likelihood between e_1 and e_2. The threshold levels mentioned above, once formulated, might provide one way of dealing with this possibility. If, for example, e_1 and e_2 have sufficiently similar probabilities, then their cost-effectiveness will be determinate of their rank in the testing process.

22. Cf. CP 7.220.

23. Note that Peirce's guidelines, which involve considerations about economy of research, speak only to the priority that experimental tests are to be accorded. They do not instruct the scientist how to devise experiments to test for the consequences of a hypothesis, nor should they be expected to do so. They will, however, inform the process by providing parameters within which the scientist operates.

REFERENCES

Achinstein, Peter (1991) *Particles and Waves: Historical Essays in the Philosophy of Science.* New York: Oxford University Press.

Achinstein, Peter (2001) *The Book of Evidence.* New York: Oxford University Press.

Earman, John (1992) *Bayes or Bust?: A Critical Examination of Bayesian Confirmation Theory.* Cambridge: The MIT Press.

Franklin, Allan (1997) "Recycling of Expertise and Instrumental Loyalty," *Philosophy of Science* 64(4): S42–S52.

Gemes, Ken (1998) "Hypothetico-Deductivism: The Current State of Play; The Criterion of Empirical Significance Endgame," *Erkenntnis* 49: 1–20.

Glymour, Clark (1980) *Theory and Evidence.* Princeton, NJ: Princeton University Press.

Lakatos, Imre (1978) *The Methodology of Scientific Research Programmes.* Cambridge: Cambridge University Press.

Mayo, Deborah G. (1996) *Error and the Growth of Experimental Knowledge.* Chicago: University of Chicago Press.

Newton-Smith, W. H. (1981) *The Rationality of Science.* London: Routledge & Kegan Paul.

Peirce, Charles S. (1931–1935) *The Collected Papers of Charles Sanders Peirce,* Vols. I–VI, Charles Hartshorne and Paul Weiss, eds. Cambridge, MA: Harvard University Press.

Peircc, Charles S. (1958) *The Collected Papers of Charles Sanders Peirce.* Vols. VII–VIII, Charles Hartshorne and Paul Weiss, eds. Cambridge, MA: Harvard University Press.

Rescher, Nicholas (1978) *Peirce's Philosophy of Science.* Notre Dame: University of Notre Dame Press.

Positive Relevance: A Defense
and a Challenge

Sherrilyn Roush and Peter Achinstein

Positive Relevance Defended, Sherrilyn Roush

The positive relevance view of evidence, in which e is evidence for h if and only if $p(h/e)$ is greater than $p(h)$—that is, if and only if e raises the probability of h above what it was when e was not taken into account—is held in high esteem by Bayesians and others who view probability as the sole concept needed to analyze the concept of evidence. One regularly hears that positive relevance is not sufficient for e to be evidence for h, since e may raise the probability of h without raising it high enough to make h as much as plausible, in which case one may not want to say that one has evidence for h. However, Peter Achinstein has objected, on the basis of putative counterexamples, that positive relevance is not even *necessary* for evidence, as a plank of his argument that probability alone cannot capture the concept of evidence (Achinstein, 1983; 2001). I will argue that these examples are ineffectual for making his point, the first because he makes a false assumption in it, and the second because what he wants us to count as evidence is redundant with evidence we are already taking into account. Without examples showing that positive relevance is not necessary for evidence, there is no reason to change to Achinstein's explanation-based view of evidence, because, as is familiar, the

The text of this chapter was previously published in *Philosophy of Science* 71 (2004): pp. 110–116 and 521–524.

purported insufficiency of positive relevance can be addressed by adding a condition requiring high posterior probability for the hypothesis.

Achinstein's first example involves a lottery:

> e_1 = *The New York Times* (NYT) reports that Bill Clinton owns all but one of the 1000 lottery tickets that exist in the lottery.
> e_2 = *The Washington Post* (WP) reports that Bill Clinton owns all but one of the 1000 lottery tickets that exist in the lottery.
> b = This is a fair lottery in which one ticket drawn at random will win.
> h = Bill Clinton will win the lottery.

e_1 and e_2 are both pieces of evidence, b is background knowledge, and h is the hypothesis. Achinstein submits that given e_1 and b, e_2 is strong evidence for h, yet he claims:

$$p(h/e_1 e_2 b) = p(h/e_1 b) = 999/1000 \text{ (Achinstein, 2001: 70)}.$$

In other words, he claims that once e_1 is incorporated as evidence, e_2 does not change the probability of the hypothesis, and therefore, on the positive relevance view, e_2 is not evidence for h in the circumstance described. Ergo, the positive relevance condition is too strong.

A little reflection shows that there is something amiss in the assignment of probabilities in this case. The probability that Clinton will win is .999 only if the probability that he owns 999 out of 1000 tickets is 1. How are we supposed to get that? The only way to get from the claim that it was reported that Clinton owns 999 tickets to the claim that the probability he will win is .999 is to assume that a report in the NYT that c (where c = Clinton owns . . .) makes the probability of c equal to 1, that is, that the NYT is a perfect transmitter. This assumption makes the probability of a Clinton win as high as it could get (.999) on the basis of any report that he owns 999 out of 1000 tickets, and thereby prevents any other such report from raising the probability. The NYT is a respectable newspaper, but this assumption is inappropriate, not just because it is false—the existence of a "corrections" section is sufficient to show this—but also because it automatically removes from consideration what goes on with probabilities when you have two imperfect sources of information, which is where all of the interest of this example lies.

It is obvious that if there exists evidence that we have and that makes the probability of a hypothesis equal to 1, then on the positive relevance view nothing else will count as evidence, because nothing can change a probability of 1. The question is whether there are any examples of that sort where the verdict strikes us as wrong. It is very hard to come up with examples where evidence makes the probability of a hypothesis equal to 1 when that probability was not already 1, that is, by raising it from a lower value, unless the hypothesis itself is taken as evidence. It is not enough simply to *assume* that a given case is one where evidence has made the probability of something, here c, equal to 1. That would have to be argued for, since it is the crucial, and difficult, premise of the argument against positive relevance. It is obvious that c in this example does not start out with probability 1, but it is also obvious that having the NYT report that c does not make $p(c) = 1$ either. What Achinstein would need to prove is implausible in this case, and that is enough to ruin this counterexample.

However, there is more to say. Even though the NYT report does not make the probability of c equal to 1, we may assume that it makes that probability high, and therefore the posterior probability of the hypothesis high. If one thinks that being evidence is a matter of positive relevance, one may also think that the degree of this relevance is proportional to the strength of the evidence, although this is not implied. That is, one may think that the strength of the evidence is measured by the degree to which it positively changes the probability of the hypothesis. If so, then the example looks strange, because it looks as if how good the NYT or WP report is as evidence on the positive relevance measure depends on whether it was discovered first, because the one discovered second has little room to change the probability of h once the other evidence has been registered. If the two reports are not independent—if, say, they both got their information from Reuters—then that seems far less strange. But the two reports could have been independent, and then there seems to be a problem. Nevertheless, this is not a reason to think positive relevance is not necessary for evidence. It is only a reason to think that the degree of relevance—measured as the difference between $p(h/e)$ and $p(h)$ or the ratio of $p(h/e)$ to $p(h)$—does not measure the degree to which one thing is evidence for another.

It is instructive to compare the likelihood ratio method as applied to this example, since this method makes it harder to slip into presuming that a report of c makes $p(c) = 1$. On this view, to determine whether e is evidence for h we compare $p(e/h)$ to $p(e/-h)$, and if the first is greater than the second, then e fulfills what a likelihood conception takes to be necessary for one thing to be evidence for another. (Fulfilling the likelihood condition implies fulfilling positive relevance.) In our case, to decide whether e_2 is evidence for h when e_1 is already in the stock of evidence, we compare $p(e_2/h.e_1)$ to $p(e_2/-h.e_1)$. $p(e_2/h.e_1)$ is clearly greater than $p(e_2/-h.e_1)$, since in the second case the given fact that Clinton does not win the lottery casts doubt on the veracity of the NYT report that he owns 999 out of 1000 of the tickets. If that report was false, then, unless we can assume that the WP always copies the NYT, e_2 has a lower probability than it has if Clinton does win the lottery and the NYT report is the same. This means that the likelihood ratio is greater than 1 if we can assume that there is some chance the WP report is independent of the NYT report, which is anyway the only case where not counting e_2 as evidence is counterintuitive.

Achinstein's second example involves an intervening cause:

e_1 = On Monday at 10 a.m. David, who has symptoms S, takes medicine M to relieve S.

e_2 = On Monday at 10:15 a.m. David takes medicine M' to relieve S.

b = Medicine M is 95% effective in relieving S within 2 hours; medicine M' is 90% effective in relieving S within 2 hours, but has fewer side effects. When taken within twenty minutes of having taken M, medicine M' completely blocks the causal efficacy of M without affecting its own.

h = David's symptoms S are relieved by noon on Monday.

In familiar form, Achinstein claims that given e_1 and b, information e_2 is strong evidence for h, because medicine M' is 90% effective in relieving symptoms S.[1] Yet the positive relevance account of evidence does not render this verdict, for:

$$p(h/e_1b) = .95$$
$$p(h/e_2.e_1b) = .90 \text{ (Achinstein, 2001: 70–71)}.$$

e_2 not only does not increase h's probability over what it was when e_2 was not taken into account, it *decreases* that probability.

In this case it is obvious that the probabilities are right. It is much less obvious that they yield counterintuitive judgments about evidence. When the example is first introduced Achinstein says we should believe that e_2 is strong evidence for h when e_1 and b are given, "since medicine M' is 90% effective in relieving symptoms S" (Achinstein, 2001: 71). However, the fact cited would justify the claim that e_2 is strong evidence for h given e_1 only if supported by one of two assumptions that are false, and which Achinstein has disavowed.

The first is that a sufficient condition for e to be evidence for h is that $p(h/e.b)$ is high. Achinstein rightly rejects this high-probability condition as insufficient for evidence, because it does not require that e be relevant to h, or, intuitively, that e "made" the probability of h high (Achinstein, 2001: 71). It would count the fact that a man consumed birth control pills as evidence that he will not get pregnant. The problem is that the probability that he would not get pregnant was already high, so his consumption of birth control pills has no work to do supporting it. The same is true of e_2 in the case of the medicines. The probability that David would recover was already high when the second medicine came along. e_2's role in making the probability that he would recover high is redundant with work that would have been done by e_1.[2] (I will have more to say later in support of this claim.) By Achinstein's own lights, the mere fact that the probability of h is high after e_2 is taken into account does not make e_2 evidence for h when e_1 is given.

89

The other assumption that would make "medicine M' is 90% effective in relieving symptoms S" a justification for the claim that e_2 is evidence for h once e_1 is taken into account is—as the quoted clause strongly suggests—that e_1 has not been taken into account! It is clear that if David had never taken medicine M, but had taken medicine M', then the fact that he had taken M' would be strong evidence that he would recover, because that medicine is 90% effective. The probabilities also conform to that judgment, since $p(h/e_2.b) > p(h/b)$. However, these are not the probabilities we are concerned with in Achinstein's claim that e_2 is evidence for h once e_1 has been taken into account, and they do not obviously bear a helpful relation to $p(h/e_1.b)$ and $p(h/e_2.e_1.b)$.

We get a different justification when Achinstein later considers the same example (with minor changes): "Isn't the fact that I am taking [M'] after having taken [M] evidence that my pain will be relieved, even though the probability that it will be relieved has decreased from 9[5]% to 90%?" (Achinstein, 2001: 84) The phrase "the fact that I am taking [M'] after having taken [M]" is notably ambiguous. The reading that invites itself is "the fact that have taken M' and taken M." It is obvious that this conjunction *is* evidence for h on intuitive grounds, but it is also clear that the corresponding positive relevance condition is fulfilled: $p(h/e_1.e_2.b) > p(h/b)$. That this conjunction is evidence is not the claim Achinstein needs to defend for his conclusion, but it is the claim the words suggest when he asks for our intuitions.

Achinstein's phrase could, and should, mean "the fact that I am taking M' *given* that I have taken M." This would mean that the question is whether, given the background and the fact that I have taken M, my having taken M′ seems like further evidence—new information—that I will recover. Consider a concerned friend who knows that I have taken M. Would she be convinced if we told her we had new evidence that I was going to recover, namely, that I had taken M′? She would undoubtedly be less confident that I would recover than she was before, if only by a little, and furthermore she would be annoyed at the misleading advertisement. It is clear that when we have e_1, and we acquire e_2, we do *have* evidence that I will recover. It is not at all clear that given e_1, e_2 is evidence that I will recover.

90

Any lingering confusion in our intuitions about this case comes, I think, from the fact that if I do recover then it will have been due to the causal efficacy of M′ alone, the medicine the taking of which is reported by e_2. In this the example differs from previous Achinstein examples of similar form, such as one in which the first piece of evidence that Freddy will win a lottery is that he owns 999 of 1000 tickets, and the second piece of evidence tallied says that by the next day 1001 tickets have been sold, of which Freddy still owns 999 (Achinstein, 1983: 152). The probability of Freddy winning is still very high after the second report is in, but it has dropped from what it was with the first report, so e_2 does not count as evidence in this context on the probabilistic relevance view. Nor should it, on the basis of what I have argued above, since e_2 is redundant with part of e_1.

One might think that although e_2 has no right to count as evidence in this case, that is because there is no sense in which someone else's buying another ticket *causes* Freddy to win if he wins. The case of the medicines is different, because if I recover then M′, and not M, will be the cause of that. The latter claim is true but of no avail, I think, because other things are true as well. For example, I would have had at least a 90% probability of recovery even if I had not taken M′. This is related to the peculiar fact that the only reason that M is not the actor in bringing about my recovery is a secondary action of M′. If I had not taken M′, then I also would not *need* M′ for recovery. M′ did not do anything relevant to whether I recover that M would not have done, supporting the claim that e_2, the report that I have taken M′, is in the most important sense redundant with part of e_1. When we learn e_2, we learn something about the mechanism of recovery, but we learn nothing new about *whether* I will recover—the point at issue in the hypothesis—except the negative news that my chances have gone down, which does not change the fact that my chances are very good.

Each of M and M′ raises the probability of recovery when acting alone, and when acting together they raise the probability over what it was with neither. This conforms to the fact that M and M′ are jointly, and each individually, evidence for my recovery. However, once I know that M has been taken, learning that M′ has been taken does not increase my confidence in recovery, and it is not evidence for that recovery because the information it gives about whether I will recover is redundant with information we already had.

A CHALLENGE TO POSITIVE RELEVANCE THEORISTS: REPLY TO ROUSH, Peter Achinstein

According to the standard positive relevance theory of evidence, a piece of information is evidence for a hypothesis if and only if it increases the probability of the hypothesis. Using a series of counterexamples, I have argued that positive relevance is neither necessary nor sufficient for evidence (Achinstein, 2001).

Sherrilyn Roush (2004) defends positive relevance as a necessary (albeit not a sufficient) condition for evidence by rejecting two of my counterexamples. In the first, let h be the hypothesis that Bill Clinton will win a certain lottery; e_1 is that the New York Times (NYT) reports that Clinton owns all but one of the tickets; e_2 is that the Washington Post (WP) reports that Clinton owns all but one of the tickets; and b is background information that this is a fair lottery of 1000 tickets, one of which will be drawn as the winner. I assume that e_2 is evidence that h, given e_1&b, even though

(1) $p(h/e_2\&e_1\&b) = p(h/e_1\&b) = 999/1000$,

that is, even though e_2 does not increase h's probability.

Roush objects to (1) because it implies that the NYT is a perfect transmitter, that is, it implies

(2) $p(e_3/e_1\&b) = 1$,

where e_3 = Bill Clinton owns all but one of the lottery tickets. She regards (2) as false, and so it is, in the real world. But let us suppose a "perfect" world in which the NYT (as well as the WP) always gets it right, so that (2) is true, as well as (1). Then my counterexample stands:

(3) Given the NYT report (e_1), the fact that the WP reports what it does (e_2) is evidence that Clinton will win (h), despite the fact that e_2 fails to raise the probability that Clinton will win, that is, despite the fact that (1) is true.

Now let's talk about the real, imperfect world in which even the NYT and the WP sometimes get it wrong. Let N = the NYT is correct in its report about how many lottery tickets Clinton holds, and let W = the WP is correct in its report about how many lottery tickets Clinton holds. I would claim that

(4) Given $W\&e_1\&N\&b$, e_2 is evidence that h.

And in this case

$$p(h/e_2\&W\&e_1\&N\&b) = p(h/W\&e_1\&N\&b) = 999/1000.$$

If so, the evidential claim (4) violates Roush's positive relevance requirement.

Both my evidential claim (3), for the perfect world, and (4), for the imperfect one, are based on the idea that in these cases the putative evidence provides a good reason to believe the hypothesis, without raising the probability of the hypothesis.

A second counterexample of mine that Roush cites is supposed to show that information can be evidence for a hypothesis even though it lowers the probability of that hypothesis. Let

e_4 = At 10 a.m. David, who has symptoms S, takes medicine M to relieve S.
e_5 = At 10:15 a.m. David takes another medicine M′ to relieve S.
b = M is 95% effective in relieving S within 2 hours; M′ is 90% effective within 1 3/4 hours, but has fewer side effects. When taken within 20 minutes of having taken M, M′ completely blocks the causal efficacy of M without affecting its own.
h = David's symptoms are relieved by noon.

My claim is that

(5) Given $e_4\&b$, information e_5 is evidence that h will be true,

since M′ is 90% effective in relieving S and its efficacy is not blocked by having already taken M. Yet

(6) $p(h/e_4\&b) = .95$, and $p(h/e_5\&e_4\&b) = .90$.

That is, h's probability is lowered by e_5, despite the fact that, given $e_4\&b$, e_5 is evidence that h (i.e., (5)). This again violates Roush's positive relevance requirement for evidence.

Roush agrees with me about the probability claims in (6), but not about (5). On her view it is the conjunction $e_5\&e_4$ that is evidence that h, given b; it is not the case that e_5 is evidence that h, given $e_4\&b$. And with the conjunction $e_5\&e_4$ as evidence that h positive relevance is satisfied, since $p(h/e_5\&e_4\&b) > p(h/b)$.

So far as I can see, the only reason Roush suggests for saying that (5) is false is that information e_5 offers "nothing new about whether [David] will recover—the point at issue in the hypothesis [h]," whereas the conjunction

e_5&e_4 does offer something new (over b alone) about David's recovery. This is reinforced by her claim at the end that knowing that e_5 is true does not increase one's "confidence" that h is true, "since the information it gives about whether [David] will recover is redundant with information we already had."

In one sense, as Roush notes, e_5 does offer something new about the recovery, namely, about its mechanism (it will be M', not M, that produces it). In another sense—that in which it provides information that changes the strength of the previous evidence—it also offers something new. Only in the sense that information e_5 does not "increase [one's] confidence in recovery" can it be deemed not new. And the latter seems to be enough for Roush to deny that e_5 is evidence that h, given e_4&b.

This is to demand of evidence that it be something that should provide a more convincing reason for believing a hypothesis than before, a reason that should increase one's confidence in the truth of the hypothesis— which, of course, fits in with Roush's idea that evidence should increase the probability of an hypothesis. In my view, which is defended in my 2001 publication and which Roush needs to confront directly, what evidence has to supply is simply a good reason for believing, not necessarily a reason that is better, or more convincing, or more confidence-producing than before (though, of course, it can do that too). Indeed, the point of my two counterexamples is that evidence can provide a good reason for belief even if (as in the first counterexample) that reason does not increase the probability of the hypothesis (and in that sense does not increase one's confidence in its truth), and even if (as in the second counterexample) that reason, while still a good one, is weaker than one we had previously, and in fact decreases the probability of the hypothesis. To reject my counterexamples, Roush and other positive relevance defenders need to show why this idea is mistaken.

ACKNOWLEDGMENT

Rice University and The Center for Philosophy of Science, University of Pittsburgh, supported the "Positive Relevance Defended" section of this chapter. Roush wishes to thank an anonymous referee for the suggestion to draw a stronger conclusion about the second example.

NOTES

1. It is not stated but must be assumed in the example that the symptoms S are such that if David did not take medicine, then he would not recover.

2. This example is analogous to another Achinstein example dealt with by Patrick Maher (1996: 172), who drew the same conclusion I have as to whether evidence is present. My reply has the advantage of not relying on Maher's particular view of confirmation, some of whose assumptions Achinstein attacked to defend himself (Achinstein, 1996).

REFERENCES

Achinstein, Peter (1983) "Concepts of Evidence." In *The Concept of Evidence*, P. Achinstein, ed. Oxford: Oxford University Press, pp. 145–74.

———. (1996) "Swimming in Evidence: A Reply to Maher." *Philosophy of Science* 63: 175–82.

———. (2001) *The Book of Evidence*. New York: Oxford University Press.

Maher, Patrick (1996) "Subjective and Objective Confirmation." *Philosophy of Science* 63: 149–74.

Roush, Sherrilyn (2004) "Positive Relevance Defended." *Philosophy of Science* 71: 110–16.

94

EVIDENCE AS PASSING SEVERE TESTS:
HIGHLY PROBABLE VERSUS HIGHLY
PROBED HYPOTHESES

Deborah G. Mayo

According to Karl Popper, "Mere supporting instances are as a rule too cheap to be worth having; . . . any support capable of carrying weight can only rest upon ingenious tests, undertaken with the aim of refuting our hypothesis, if it can be refuted" (Popper, 1983: 30). As intuitively plausible as this remark is, philosophical accounts of evidence, including Popper's, have been stymied by the problem of identifying those cases where accordance between data x and a hypothesis H should really count as evidence for H. As increasingly available formal techniques for data mining and model selection bring the price of obtaining "good fitting models" ever lower, the problem is further exacerbated. The latest high-powered computer packages offer a welter of algorithms for "automatically" selecting among this explosion of models, but as each boasts different, and incompatible, selection criteria, we are thrown back to our basic question: What is required to severely discriminate among well-fitting models such that, *when a claim (or hypotheses or model) survives this test the resulting data count as good evidence for the claim's correctness, or dependability, or adequacy*. The question, as I see it, is fundamentally one for philosophy of science, although its adequate answer demands a combination of methodological, empirical, and statistical means.

In this chapter I sketch a conception of evidence as the result of surviving a severe or risky test:

> *Data x in test T provide good evidence for inferring H (just) to the extent that hypothesis H has passed a severe test T with x.*

(I use the word *hypothesis* to cover claims about models as well as the more usual statistical hypotheses.) This discussion is intended as a continuation and extension of the existing program that I call "the error statistical account," being built on ideas from the school of statistical testing based on frequentist error probabilities (e.g., significance levels, confidence levels). On this conception of evidence, even claims that are highly beliefworthy, or highly probable (however probability is construed), would not get credit from tests that poorly probed them; evidential credit goes only to those specific hypotheses or models that have survived probative testing.

I begin by explicating the concepts involved more clearly than before to address several misunderstandings, both of the statistical tools and of the broader account of evidence and inference that I wish to base upon those tools. I then confront a general type of criticism that is repeatedly raised (usually, but not exclusively from Bayesian quarters) which would, if correct, vitiate the account of evidence I am advancing, namely, that *the severity account of evidence is inadequate because hypotheses that pass severe tests on my account may be accorded low posterior degrees of support, probability, or belief.* I consider some of the variants of the criticism raised by philosophers, such as Achinstein and Howson, as well as by statistical practitioners such as Berger, Cohen, and Meehl. The problem revolves around what might be dubbed the *"highly probed versus highly probable debate,"* and although it has been around for some time, progress has been greatly hampered due to the lack of an adequate account of "highly probed" (i.e., of passing a severe test). After clarifying the notion of severity I have in mind, I argue, with respect to each variant of the criticism, that (1) the probabilistic assignment commits a fallacy, which may be called the *fallacy of instantiating probabilities*; and (2) the error statistical assessment of the data, but not the assessment advocated by the critic, is in sync with the goals of severity, as well as our intuitions about when data should count as supporting evidence in science.

The Severity Requirement: Some Examples

College Readiness

If we are testing how well a student, Isaac, has mastered high school material so as to be considered sufficiently ready for work in a four-year college, a test that covers work from 11th and 12th grade science, history, and mathematical problems (in geometry, algebra, trigonometry, and pre-calculus), requires writing a critical essay, and so on, is obviously *more difficult to pass* than one which only requires showing minimal proficiency in these subjects at a 6th or 7th grade level. It would be regarded as more searching, more probing, and more severe. The understanding behind this commonplace judgment is roughly this: Achieving a passing or high score is easier and more likely to have come about with the less severe test than the more severe one, *even among students who have not mastered the bulk of high school material, and hence are not "college-ready."* We deny that Isaac's passing score provides good evidence that he fully masters high school material if we learn that even students who mastered little of the material would very probably have scored as high as, or even higher than, he did. In so doing, we are demanding that *before regarding a passing result as genuine evidence for the correctness of a given claim or hypothesis H, it does not suffice to merely survive a test; such survival must be something that is very difficult to achieve if in fact H deviates from what is truly the case.*

By the same token, if the test *is* sufficiently stringent, such that it is practically impossible for students who have not mastered at least p% of high school material to achieve a score as high as Isaac's, then we regard his passing grade as evidence that he has mastered at least this much. The same reasoning abounds in science, as some recent examples illustrate.

97

Hormone Replacement Therapy

Until very recently, millions of women in the United States were routinely prescribed hormone replacement therapy (HRT) on the grounds that there was excellent evidence that:

> H: HRT is advantageous to menopausal and post-menopausal women.

Data, collected over many years, were taken to show HRT's benefits, not only for the effects of menopause, but also for reducing the risk of heart attack and stroke, various cancers, memory loss, and a host of age-related deficiencies—thereby allowing women to remain "Feminine Forever," to cite the title of one famous book.[1] The results of recent, large, randomized treatment-control studies, however, are widely taken to show that such data actually failed to provide evidence for the alleged benefits. Moreover, the new data are regarded as strong evidence that HRT, especially if taken for more than five years, increases the chance of breast cancer and fails to provide the supposed protection against heart disease and memory loss.

"We painted too rosy a picture," the acting director of the Women's Health Initiative conceded.[2] Due to the nature of the retrospective studies on which such sanguine recommendations were based, the previous studies are now known to have had little capacity to distinguish the benefits of HRT from confounding factors. In particular, the women using HRT have been found to have characteristics that are separately correlated with the beneficial outcomes (e.g., they are healthier, have better access to medical care, and are better educated than women not taking HRT).

The underlying reasoning is this: *data x do not supply evidence for a (causal) hypothesis if it would have been just as easy for the observed association between x and the hypothesized factor in H to have come about due to factors other than the one claimed in H.* The agreement (between *x* and *H*) from the retrospective studies failed to provide evidence in support of *H* because *H* did not thereby pass a test that we would consider severe.

The Columbia Space Shuttle

98

Within hours of the Columbia space shuttle disaster, it was hypothesized that the crash was due to overheating from foam tiles striking the left wing, and yet the available data behind this early conjecture (temperature readings, photographs of displaced tiles during takeoff) was scarcely solid evidence for this hypothesis. Although the "damaged tile on ascent" hypothesis seemed to fit, and could "account for" the data, the early data did not constitute the results of probing numerous other possible fault lines.

Contrast this to the recently completed report by the NAS Board. First, the NAS report tightens up the fit by providing a very detailed version of the "damaged tile on ascent" hypothesis:

> *H*: The physical cause of the loss of *Columbia* and its crew was a breach in the Thermal Protection System on the leading edge of the left wing, caused by a piece of insulating foam which separated from the left bipod ramp section of the External Tank at 81.7 seconds after launch, and struck the wing in the vicinity of the lower half of Reinforced Carbon-Carbon panel number 8.

Second, the report presents "aerodynamic, thermodynamic, sensor data timeline, debris reconstruction, and imaging evidence—to show that all five independently arrive at the same conclusion," *H*, which has thereby passed a severe test. *When we evaluate evidence in this way, we are scrutinizing inferences according to the severity of the tests they have survived.* Note that the severity assessment, as is typical, is based not on a single data set but on numerous separate probes taken together.

SEVERE TESTS

From these examples, we can identify what is required for a hypothesis (or model or prediction) *H* to have passed a good or severe test with data *x*. Although at minimum a passing result requires a "good agreement" between data *x* and hypothesis *H*, more is required. In addition to finding a good fit between *x* and *H*, we need to be able to say that the test was really probative—that so good a fit between data *x* and *H* is practically impossible or extremely improbable (or an extraordinary coincidence, or the like) if in fact it is a mistake to regard *x* as evidence for *H*.

Severity Requirement

So one way we can encapsulate the severity requirement is this:

> *Hypothesis H passes a severe test T with x if (and only if):*
> (i) *x* agrees with or "fits" *H* (for a suitable notion of fit[3]), and
> (ii) test *T* would (with very high probability) have produced a result that fits *H* less well than *x* does, if *H* were false or incorrect.

Data *x* provide good evidence for inferring *H* only if *x* results from a procedure which, *taken as a whole*, constitutes *H* having passed a severe test—that is, a procedure which would have (at least with very high probability) unearthed any error or flaw in the inference to *H*. Far from wishing to justify the familiar inductive rule from observing that n% of A's have been B's in a sample, to an inference that n% of A's are B's in a given population, we can see that such a rule would license inferences that had not passed severe tests, and would be highly unreliable. Likewise we eschew going from correlational data to causal claims, without demonstrating that errors have been well ruled out. Correlational data *x* provide good evidence for *H* only when *x* is generated in a procedure that did a good job of ruling out the ways it can be an error to go from *x* to *H*.

99

Qualitative Assessments of Severity

Although the previous section encapsulates severity in terms of probability, it is not required that a formal probability model be definable to sustain a severe test; in fact, the strongest severity arguments seem to arise from wholly informal appraisals. For instance, were Isaac to continually score high marks, regardless of how difficult or advanced we make the exam (SAT, advanced placement tests, etc.), we are clearly on firm ground in regarding the passing results as excellent evidence that Isaac knows the given material without any recourse to formal models. The term "probability" here may serve merely to pay obeisance to the fact that all empirical claims are strictly fallible, even if a counterexample is never actually instantiated. In engineering and other contexts it is also common to work without a well-specified probability model for catastrophic events, and yet the same requirement about evidence holds. Modifying my definition for such contexts, the engineer Yakov Ben-Haim suggests, "We are subjecting a proposition to a severe test if an erroneous inference concerning the truth of the proposition can result only under extraordinary circumstances" (Ben-Haim, 2001: 214).[4] If one thinks of the circumstances that would have to obtain for the NAS board to be substantially wrong in their interpretation of the data, one gets the idea of what is meant by "extraordinary" here.

Severity Principle

Whether severity is understood quantitatively or qualitatively, in terms of probability or in terms of non-probabilistic, but still formal, notions, the over-arching principle of evidence remains. We may refer to it as *the severity principle*:

> *Severity Principle*: Data x (produced by process G) provide a good indication or evidence for hypothesis H (just) to the extent that test T severely passes H with x.

Within this analysis, hypothesis H is regarded (or modeled) as a claim about some aspect of the process that generated the data, G. According to the severity principle, when hypothesis H has passed a highly severe test (something that may require several individual tests taken together), we can regard data x as supplying good grounds that we have ruled out the ways it can be a mistake to regard x as having been generated by the procedure described by H.

DANGEROUS MISUNDERSTANDINGS

Although a full understanding of the severity principle, and of how to calculate severity, demands careful discussion beyond this paper, the central points I need to make require avoiding some common misunderstandings, especially in regard to the context where the severity requirement refers to a probabilistic model.

A Test (in the current account) Does Not Require Starting With a Hypothesis H

I find it useful to adopt the language of testing because it seems the best way to highlight the challenge (or *agon*) that an inference is required to survive before allowing that there is evidence for it. However, there are common conceptions philosophers typically hold about tests that I wish to deny (C. S. Peirce may be the sole exception). In particular, there is no presumption that the hypothesis is arrived at first (somehow), that is, it is not assumed that the data x must be "novel" in some sense. Beginning with data x and appropriately arriving at a hypothesis H (which might be a model, or other claim), as in cases of estimation or model searching, may still permit x to be a severe test for H. In fact, I developed the severity notion precisely to distinguish between legitimate and illegitimate cases where data x are used both to arrive at and test a hypothesis, that is, where a hypothesis H is "use-constructed" (see, for example, Mayo, 1991; 1996).

What our analysis demands is recognizing *how* various data-dependent procedures *may* create obstacles for the severity with which given claims may be said to pass (analogous to the way significance levels and other error probabilities are altered by certain selection procedures and stopping rules in statistics).[5] However, so long as the overall analysis satisfies severity requirements, x is good evidence for H (or, equivalently, H has passed a severe test with x).

A Severity Assessment Is Always Relative to the Hypothesis That "Passes"

It is common to talk as if a severity assessment attaches to the test itself—but doing so leads to untoward results. One cannot answer the question: How severe is test *T*? without including the particular inference that is claimed to have passed the test, if any. In other words, whatever claim or hypothesis one is contemplating inferring, on the basis of x, is the claim to regard as *passing* for purposes of appraising severity. The great advantage of relativizing the assessment to the particular inference (and the particular data set) is that high severity is always what is wanted for evidence.[6] No problem occurs unless one forgets that a given test may severely pass one hypothesis and not another, even among the hypotheses under consideration. This confusion most readily takes the form of what might be called *the criticism from overly sensitive tests*.

The Criticism from Overly Sensitive Tests

Severity cannot be a sensible desiderata, so the criticism goes, because a test may be made so severe that even a trivially small departure from a hypothesis H will result in inferring H'—where H' is a rival to H, or an assertion about some anomaly or error in H. What this criticism overlooks is that the inference whose severity we would need to consider in that case is H'; but having put H to a stringent test is not to have stringently probed H'! The misunderstanding behind the criticism boils down to thinking that H' has passed a severe test, as I am defining it, but in fact it is quite the opposite.

Consider our test for deficiencies in college readiness and the hypothesis:

> H: Isaac is college-ready (i.e., not deficient) ,

as against

> H': Isaac is not college-ready.

We can make the tests so hard, and the hurdle for regarding grades as evidence for H so high, that his scores are practically always going to lead to denying H and inferring H' (he is deficient). However, H' has passed a test with *very low* severity because it would very often lead to inferring H', even if H' is false and actually H is true.

The Criticism from Overly Sensitive Tests is often given with regard to testing models: Look, if we make the test so severe, the critic says, we are always going to find some flaw in the model—models are always approximate. This not only fails to employ "severity" as I have defined it, but it also over-looks the central job it is designed to perform: (a) Severity lets us show that if we make the test so sensitive, then the assertion "this model is flawed" is *not* going to pass a severe test; and (b) a severity assessment issues in a report of what (if anything) *does* pass severely (e.g., in statistical tests, it reveals just how small the discrepancy indicated is).[7] (In a severity interpretation of statistical tests, (b) is formalized in a "rule for rejection." See Mayo, 1996.)

Severity Condition (ii) Differs from Saying That x Is Very Improbable Given Not-H

That is, condition (ii) is not merely to assert that $P(x;H$ is false$)$ is low, where "$P(x;H$ is false$)$" is to be read: "the probability of x under the assumption that H is false."[8] This is called the *likelihood* of H given x. For a familiar example, H_1 might be that a coin is fair, and x the result of n flips. For any x one can construct a hypothesis H_2 that makes the data maximally likely; for example, H_2 can assert that the probability of heads is 1 just on those tosses that yield heads, and 0 otherwise. $P(x;H_1)$ is very low and $P(x;H_2)$ is high, but H_2 has not passed a severe test because one can always construct *some such maximally likely hypothesis or other* to perfectly fit the data on coin tosses, even though it is false and the coin is perfectly fair (i.e., H_1 is true). The test that H_2 passes has minimal severity. (This is a case of what I call "gellerization.")

A remark on notation: I am using ";" in writing $P(x;H)$—in contrast to the notation typically used for a conditional probability, $P(x/H)$—in order to emphasize that severity does *not* use a conditional probability which, strictly speaking, requires that the prior probabilities $P(H_1)$ be well-defined, for an exhaustive set of hypotheses. As we will see, such priors are not well defined in the frequentist severity account within which I am working.

The Degree of Severity with Which a Test Passes H Is Not the Degree of Probability of H

Finding that a hypothesis H severely passes test T with data x does not license a posterior probability assignment to H, a notion which depends on having prior probability assignments to the hypotheses under consideration. Such Bayesian calculations (from one of a number of schools of Bayesianism) are at odds with the severity principle in general. This, of course, is behind the "highly probed versus highly probable" subtitle of this chapter, and it is an issue to be unpacked in detail in later sections.

It should be emphasized that "H is false" is not the so-called *catchall factor*, that is, the disjunction of hypotheses other than H, including those not yet even thought of. Instead, it refers to a specific error that hypothesis H may be seen to be denying. The particular experimental context serves to ensure that H is sufficiently local so that H and its complement exhaust the space of hypotheses for the experiment at hand.

103

Logics of Evidential Relationship: Subjective Bayesian Philosophy

The severity account of evidence is at odds with familiar philosophical accounts of evidence or confirmation that seek to provide one or more measures of the logical relationship between given evidence (or evidence statements) and hypotheses. A leading example of such a *logic of evidential relationship* (E-R logic) is the subjective Bayesian account, which I understand here to refer to the kind of simple model found in Howson and Urbach (1989; 1993) and in many other philosophical works. "Inductive logic—which is how we regard the subjective Bayesian theory—is a theory of (degree of belief) consistency and thereby also a theory of inference from some exogenously given data and prior distribution of belief to a posterior distribution" (1993: 419). In the Bayesian model, data x would usually be regarded as evidence for H to the extent that the posterior probability in H given x exceeds the prior in H, although various measures of extent of evidence are possible.

The reliance on subjective probabilities in an exhaustive set of hypotheses; the fact that the import of the data comes in only by way of the likelihoods; and that "how you came to accept the truth of the evidence, and whether you are correct in accepting it as true" are regarded as "simply irrelevant" to the account are all reasons that the Bayesian approach is at serious odds with the conception of induction as severe testing (ibid., 1993: 407). Strong degree of belief in a hypothesis H, coupled with believing that the data x are very improbable under alternatives to H, suffices for high Bayesian support for H. It does not suffice for regarding H as having passed a severe test.

Existing quantitative E-R measures of support, confirmation, or the like may be evaluated by means of the severity requirement by regarding each as supplying one or another measures of "fit" (as in condition (i) of severity). However, for any such measure of fit between x and H, the severe tester wants to ask: *How often would so good a fit result (from a series of applications of the test) under the assumption that H is false?* The probability that a hypothesis *H* would pass a test, under the assumption that *H* is false, is an *error probability* or error frequency. The severity with which *H* passes is high just in case (or just to the extent that) this error probability is very low.[9] It is this crucial role of error probabilities that makes it appropriate to call the severity account an error probability account, even where the assessment remains qualitative. Because the error probability assessment demands taking this additional step beyond the Bayesian or other evidential-relation accounts, it is not surprising that the latter fail to control error probabilities, as we have defined them.

POPPERIAN SEVERITY

It was Popper who is best known for insisting on severe tests, and while the severity requirement is clearly Popperian *in spirit*, Popper never adequately captured the severity notion. Although Popper offered various formal definitions that would *potentially* measure the degree to which *x* corroborates *H*, $C(H,x)$, he claimed that in order for it to actually measure corroboration, *x* would have to be the result of a severe test: "In opposition to [the] inductivist attitude, *I assert that $C(H,x)$ must not be interpreted as the degree of corroboration of H* by *x*, unless *x* reports the results of our sincere efforts to overthrow *H*. The requirement of sincerity cannot be formalized—no more than the inductivist requirement that *x* must represent our total observational knowledge" (Popper, 1959: 418; I substitute his *h, e* with *x, H* for consistency with my notation).

Under "inductivist" Popper includes Bayesians both of Carnapian and subjective varieties, as well as those holding variants of induction by enumeration. The important kernel of rightness here is that these inductive logics of evidential relationship made it too easy to find evidence in support of hypotheses: *Their measures of evidential relationship may be satisfied without satisfying the requirement of severity.* The formal counterpart to this claim is that all such algorithms lack formal niches through which to pick up on aspects of the evidence that are relevant for assessing if the test was really good at probing *H*'s errors, that is, niches for assessing severity in my sense. For example, there are aspects of the generation of data and the specification of hypotheses to test that alter a test's error probabilities and yet do not change the ratio of likelihoods of the hypotheses. Hence, if "same likelihood" means "same evidence" there will be no way to formally pick up on a difference that makes a difference, at least to one who requires severity. Unfortunately, Popper's computations suffered from just this weakness.[10]

Popper's Comparativist Account

According to Popper, data x pass H severely if: H fits or entails data x and x is improbable "without" H or under the assumption that H is false. But his formal measures never got beyond requirements about comparative likelihoods, such as

(a) $P(x;H)$ = high (or maximal)

(b) $P(x;H$ is false) = low

Moreover, Popper had no machinery to compute requirement (b), but instead supposed (b) is satisfied when

(b)' $P(x;H')$ = low

where H' is the currently best-tested alternative. However, since H and H' need not be, and generally would not be, exhaustive of the space of possible hypotheses, (b)' certainly does not warrant (b). Data may satisfy Popper's requirement without satisfying severity as I am using that term; the mere fact that x is counterpredicted by the currently best-tested alternative to H does not suffice for severely probing H. Moreover, even (b), as Popper states it, seems to be merely a likelihood, and not an error probability.

A Simple Comparative Likelihood Account

Popper's definition boils down to what may be called *a simple comparative like-lihood account*: x is evidence for H if the likelihood of H exceeds the likelihood of alternative H'. (Hacking 1965 had at one time embraced such an account.) Of course this clearly does not do justice to what Popper sought to require, which seems to be the reason he despaired of formally characterizing the conception that he held to be so vital. Popper never saw how error probabilistic notions permit capturing the intended severity requirement rigorously, rather than leaving it at the vague level of the psychological intentions of the tester (to sincerely try to find flaws in H).[11]

An equally, if not more crucial difference between the approach I am putting forward and Popper's is that even where a hypothesis has passed a test that is severe by Popper's lights—even if it is highly *corroborated*—he regarded this as at most a report of the hypothesis' past performance and denied it afforded positive evidence for its correctness or reliability. In contrast, according to the severity principle, when a hypothesis H has passed a highly severe test we can infer that the data x provide good evidence for the correctness of H.

What Should the Role of Formal Statistical Ideas Be in an Account of Evidence?

Chastened by the failures of purely formal, context-free, inductive logics, many philosophers of science nowadays tend to be critical of appealing to formal statistical ideas—Bayesian or frequentist (error statistical)—in erecting an account of evidence. Although we can agree with their doubts about purely formal E-R logics, failure to avail themselves of the methods and models of current statistical inference and modeling in science has been a major obstacle to developing philosophies of inference that are relevant to understanding and solving problems about evidence, both in science and in philosophy. In rejecting appeals to statistical methods as tantamount to advocating uniform, content-free approaches, philosophers of science have too readily given up on being able to say anything that is both general and relevant to the actual problems of evidence.

106

Are Statistical Methods Relevant Only to Formal Statistical Practice?

A related, and also mistaken, stance one often hears as grounds to reject, or at least minimize the relevance of, appeals to statistical methods in developing accounts of evidence and inference, is the supposition that such appeals could only be relevant for scientific inferences that explicitly make use of formal statistical ideas. But scientists evaluated evidence, the objection continues, before the development of statistical tools, and even now do not necessarily appeal to them (e.g., Chalmers, 2001). The flaw behind such objections is that they overlook the main philosophical goals behind appealing to statistical ideas; namely, to capture enough of the ingredients of scientific reasoning to solve philosophical problems about evidence and inference (Duhem's problem, underdetermination, theory-ladenness), and to understand and scrutinize various strategies and methodologies (e.g., preferring novel predictions, varying evidence, replication).

The Dean's Challenge

Peter Achinstein, a philosopher of science with whom I so often agree, likewise parts company with me when it comes to my advocating an appeal to statistics in building a philosophy of evidence. Achinstein concedes, in a pessimistic response to his dean's challenge as to the relevance of philosophy for science, that "standard philosophical theories about evidence are (and ought to be) ignored by scientists" (2001: 3). On the one hand, he is correct to declare philosophical accounts of evidence irrelevant to scientists, if those accounts view the question of whether data x provides evidence for H as a matter of purely logical computation. Whether data provide evidence for hypotheses, Achinstein rightly insists, is not an a priori but rather an empirical matter, and Achinstein, perhaps more than any other philosopher of science, has identified the inadequacies and counterexamples in existing E-R accounts.

On the other hand, one may reach a very different position from him as to what follows from such failures, and here is where we differ. He appears to take those failures to show that appealing to statistical ideas cannot provide the basis for a successful philosophical account of evidence. If the question of evidence is an empirical scientific matter, then in Achinstein's view philosophers can best see their job as delineating the concepts of evidence that scientists seem to use, leaving it up to scientists to apply them. However enlightening such conceptual analysis may prove to be, I think it is a mistake to curtail the philosopher's job in this fashion, and this mistake rests on the erroneous assumption that statistical accounts of evidence are restricted to supplying purely formal probabilistic computations. On the contrary, statistical ideas and methods—as used in practice—provide just the right blend of empirical and formal tools to provide forward-looking methods for appraising evidence. They can and should be at the heart of philosophical discussions aimed at resolving controversies about evidence in science—although we can grant that they generally have not been.

High Posterior Probability versus High Severity

The shortcomings Achinstein finds in probabilistic accounts of evidence serve to highlight key weaknesses in familiar Bayesian approaches: An increase in the posterior probability of H given x does not suffice for x to be evidence for H, even if coupled with his requirement that the posterior, $P(H/x)$, be high. However, by relying on such Bayesian assignments to locate the role for probability, Achinstein concludes—prematurely, in my opinion—that all statistical accounts fail. In particular, while he takes a necessary condition for x to be evidence for H that $P(H/x)$ be high, for some "objective" notion of probability, he requires also that there be the right kind of explanatory connection between x and H—something that he regards as going beyond any statistical assessments.

But why suppose that the work statistics can do for us is limited to such probabilistic computations? Rather than reason, as Achinstein seems to, that since the high posterior probability requirement fails to provide a sufficient condition for evidence, it follows that statistics cannot supply an adequate account of evidence, one may instead reject this requirement (which, after all, does not even enter into the assessment of error probabilities) and appeal to statistics to capture the severity requirement for evidence. That is the path I follow.

This appeal to statistics most nearly finds its home in non-Bayesian or frequentist accounts of statistics encompassing significance tests, (Neyman-Pearson) hypothesis tests, and estimation methods, as well as newer additions to this group of (error statistical) tools. In these accounts we find statistical methods that lend themselves to a conception of tests wherein probability arises, not to quantify support or probability in hypotheses, but to assess the probativeness (or reliability, or "trustworthiness") of the overall test procedure. From such accounts we learn at once that a methodology for severe testing cannot begin with "given" statements of evidence, but requires enough information about how the data were generated, and about the specific testing context, to assess the overall severity with which a claim or hypothesis may be inferred.

Warranting the Data

Given that "statistics claims to deal in a broad way with the collection and analysis of data" (Cox, 1981: 289), it should not be surprising that it would contain clues for a philosophical account of evidence. In particular, by viewing severity as the goal, we begin to see how statistical methods of data collection and analysis may be appealed to in order to get around what philosophers have typically considered overwhelming obstacles to justifying ampliative inference. While it is true that intermediary inferences are often required to arrive at inductive evidence, far from posing a threat to reliability, as is often thought, they may become the source of avoiding these very threats. Statistical ideas teach us how, by appropriately combining data, we may arrive at highly reliable claims from highly shaky data. The statistical cases encapsulate the reasoning in more qualitative contexts, as when the reports on the Columbia crash built inference upon inference to arrive at a clear pinpointing of blame.

A Severity (Re)interpretation of Neyman-Pearson (and Related) Methods

While retaining the central feature of these accounts, the more general "error-statistical" rubric frees us to reinterpret standard statistical accounts so as to avoid ever-present criticisms and misuses. In particular, we can reject the view of statistical hypotheses tests (e.g., Neyman-Pearson N-P tests) as mechanical tools with low long-run error probabilities, and construe them instead as tools for obtaining reliable experimental knowledge and severe tests.[12] The appeal to statistics in a philosophy of evidence, as I see it, is a two-way street: it gives insights into philosophical problems and confusions, while at the same time it serves to avoid fallacies and resolve debates in statistics. How to reinterpret N-P tests so that they supply post-data severity assessments that are sensitive to the actual outcome (unlike standard type I and type II error probabilities), and how this avoids recalcitrant problems, is discussed in detail elsewhere (Mayo, 1983, 1985, 1996, 2002b, c; Mayo and Spanos, 2000).

My concern in the remainder of this chapter is to reply to variations on a single type of criticism that has dogged non-Bayesian accounts, both in philosophy and statistical practice, namely, *that error statistical tests do not give us what we really want from an account of evidence, because they may regard x as good evidence for H, even though H is not accorded a high posterior probability*, according to one or another recommended way to obtain the requisite priors. This more general criticism can be and has been used as ammunition for a criticism directed specifically at the severity requirement, namely, that a hypothesis may have passed a severe test (it may be highly probed) even though it is not accorded high probability. The challenge revolves around what I have dubbed the *"highly probed versus highly probable debate."*

After clarifying the notion of severity I have in mind, I argue, with respect to each variant of the criticism, that (1) the probabilistic assignment commits a fallacy, which may be called the *fallacy of instantiating probabilities*, and (2) the error statistical assessment of the data, but not the assessment advocated by the critic, is in sync with the goals of severity, and with our intuitions about when data should count as supporting evidence in science.

ACHINSTEIN'S CRITICISM OF THE SEVERITY ACCOUNT OF EVIDENCE

I will begin with a simple variant on this criticism, as articulated by Peter Achinstein, and then develop and strengthen his charge in order to respond to the strongest Bayesian criticism in recent statistical literature.

We can get to the heart of the problem in short order: Achinstein's criticism assumes that (an appropriately random) sample resulting in 40% A's being B's suffices to pass, with severity, the statistical hypothesis that the population proportion of A's that are B's is .4—but this is false, at least in his example. Perhaps he is assuming that severity in the error statistical account is captured by what we termed the "simple comparative likelihood account," and admittedly, we said, Popper's view is open to such a reading. But this overlooks, and is at odds with, the central tenet of the severity account: *good fits (whether absolute or comparative) alone do not suffice for good tests!* Examining Achinstein's example and criticism serve to both illustrate the error statistical approach and set the stage for identifying key flaws in a whole cluster of Bayesian criticisms that run to this type.

Binomial Test T_1

Let us call the test he describes test T_1. A random sample of size n = 100 is taken, $\mathbf{X} = (X_1, \ldots, X_n)$, where each X_i is distributed as a Bernoulli random variable with unknown mean p, the probability of success, where in this case "success" means drawing a white ball in a random selection (he assumes that p is constant and trials are independent). We are to test two *simple* or *point* statistical hypotheses:

$$H_0\!: p = .4 \qquad \text{vs.} \qquad H_1\!: p = .6$$

Test T_1 may be described as a "point against point" test, since H_0 and H_1 each asserts just one of the possible parameter values. (H_0 is often called the "null" hypothesis, though none of my points turn on which one we regard as the null.) Such point vs. point (or simple against simple) tests are highly artificial and, strictly speaking, are not proper Neyman-Pearson tests because the hypotheses do not exhaust the parameter space, which includes all values from 0 to 1. However, because this is automatically taken into account in applying the severity criterion, we can proceed to the problem that Achinstein claims T_1 poses for my account.

Basic Concepts: Test Statistics, Significance Levels, Tail Areas

A statistical test is defined in terms of a *test statistic* or *distance measure* $d(\mathbf{X})$. In this example:

$$d(\mathbf{X}) = (\bar{X} - p)/\sigma_x,$$

where $\bar{X}$ is the sample mean, and the sample standard deviation

$$\sigma_x = \sqrt{\frac{p(1-p)}{n}}$$

is about .05.[13] The outcome is given as $\bar{X}_{obs} = .4$.

To get the *significance level* of the observed difference, we ask: "How improbable is it to observe an $\bar{X}$ as far or farther from the value hypothesized in H_0, if in fact H_0 is true?"

To answer this we must calculate $P(\bar{X} > .4; p = .4) = P(d(\mathbf{X}) > 0) = .5$. Since .5 is not small, this would yield a *nonstatistically* significant difference, so the test does *not* reject H_0. (Typically, it would be required that the statistical significance reach values as small as .05, or .01, corresponding to observing at least 50% or 55% white balls in this test.)

Note that calculating the significance level requires calculating not just the probability of the data point (.4) under H_0, but also the "tail area"— that is, the probability of outcomes *beyond* .4— under H_0. (It makes little difference in this test whether we consider > or ≥.)

The Criticism: First Variant

According to Achinstein, "On [Mayo's] view, . . . the result . . . (40 of the 100 balls selected are white) is good evidence for the hypothesis" H_0 (Achinstein, 2001: 134), because he supposes that I would regard H_0 as surviving a severe test. But has H_0 passed a severe test with outcome $\bar{X}_{obs} = .4$? No!

III

Let us abbreviate the severity with which hypothesis H passes test T_1 with outcome $\bar{X}_{obs}$ as:

$$SEV(H, \bar{X}_{obs}, T_1).$$

Although we can allow that $\bar{X}_{obs} = .4$ *fits* H_0, to calculate $SEV(H_0, \bar{X} = .4, T_1)$ requires calculating P(a "worse fit" with H_0; H_0 is false) ~ .5—and this is clearly not a high severity value! If one were to take this outcome as grounds for accepting H_0 we would erroneously do so 50% of the time! [*Note*: Since this is a non-significant result, the severity assessment happens to equal the observed significance level (or P-value), that is, .5.] To put this in other words, we may agree that H_0 would not be rejected by this outcome—but this is not tantamount to finding evidence *for* H_0. Indeed, taking no evidence against the null as evidence for it is a well-known fallacy.

To explain why, note that "H_0 is false" is the disjunction of values of p other than .4. Since the alternative statistical hypothesis in Achinstein's example, H_1, is in the positive direction, "H_0 is false" would generally be regarded as the one-sided alternative, p > .4. (The same argument can be made out if it is a two-sided alternative.) We cannot regard a failure to reject a null, that p = .4, as grounds for H_0: p = .4—that there is 0 discrepancy from .4—because the test would very probably have failed to reject H_0, even if in fact there *are* discrepancies from .4.

Suppose, for example, that the true proportion, p = .41. Of course we do not know the true value of p, but we need to consider the properties of our test under such hypothetical scenarios in order to ascertain what has and has not passed a test with severity. How severely can we say the data have ruled out a discrepancy of .01? In other words, What's the severity of the test that the hypothesis p < .41 may be said to have passed? We calculate

$$SEV(p < .41, \bar{X}_{obs} = .4, T_1) = P(\bar{X} > .4; p = .41) = P(D > -.2) \sim .6$$

But .6 is not very severe—40% of the time the test would fail to reject H_0 even if the underlying true value of p were .41. (We would typically want a severity of at least .9 before counting it as severe, although rather than select a single cut-off point, one may just report the severity obtained.) So we surely cannot say we have ruled out the existence of *any* positive discrepancy from .4, but without this we cannot say that the null hypothesis H_0 has passed a severe test. Therefore, we cannot regard such a result as good evidence for .4.[14]

However, we can find discrepancies that *are* ruled out with severity, and it is informative to calculate several, or even the entire severity curve.[15] Here are just a few:

$$SEV(p < .5, \bar{X} = .4, T_1) \sim .97$$

$$SEV(p < .6, \bar{X} = .4, T_1) \sim .999$$

So we *would* be entitled to rule out with severity that the true value of p was as great as .6, that is, infer p < .6 (so *x* is evidence that p < .6 and also that p < .5). But Achinstein's alternative hypothesis H_1 asserts p = .6, so we seem to be agreeing that our result is evidence against H_1! It is, but this does not constitute evidence for the hypothesis H_0. Fallacies stemming from applying tests to non-exhaustive hypotheses are legion (Mayo and Spanos, 2004).

The Criticism: A Bayesian Variant

Having agreed that our result is evidence against H_1, that p is as large as .6, we can mount the kind of criticism that Achinstein wishes to consider. It goes like this: For any hypothesis H that an error statistician such as myself regards as having passed a severe test, we can imagine that not-H has a high enough prior probability so that $P(\text{not-}H/x)$ is high, and thus $P(H/x)$ is low. Therefore, I would be claiming that there is evidence for H even though the posterior probability for H is low. (*Note*: Here we use the conditional probability symbol "/", since Achinstein is mounting a Bayesian criticism.)

To help Achinstein's criticism along as much as possible, let us put the evidential claim that we concurred with in a positive form: We have evidence that H': p < .6. H' is a familiar complex statistical hypothesis, in contrast to the simple point hypothesis; it includes a disjunction of values. Now for the frequentist, we said, the hypothesis H is an assertion about the data-generating procedure (G) that gave rise to the observed data. Any such hypothesis would be regarded as correct or incorrect, even though these assertions can, and generally would, be qualified in terms of how good an approximation we have evidence for, for example, by reporting a margin of error. Doing so, however, would *not* be to assign a probability to any particular statistical hypothesis H.

The entire error statistical approach is deliberately designed to avoid appealing to priors, which are nearly always unavailable or irrelevant for scientific contexts. But the critic will proceed by trying to identify a prior probability for hypothesis H that even a frequentist can, allegedly, condone.

A Prior That Even a Frequentist Can Love?

Achinstein, to his credit, is seeking an objective account, and like many others, he assumes that the way to get objective prior probabilities is through relative frequencies of certain sorts. There are two or three main gambits used to obtain priors that are allegedly "kosher for frequentists," and we will consider them in turn. The first, the one to which Achinstein appeals, requires us to consider the population or data-generating procedure G, about which the statistical hypotheses make assertions, as itself selected from a population of populations. It is assumed that: *if we randomly select a population (i.e., a bag) from a population of populations (population of bags), p% of which have hypothesis H true of them, then the frequentist prior probability of H is p.*

113

This assumption, although plausible-sounding, is fallacious. Suppose our high school student, Isaac, has been randomly selected from a population of high school seniors, 10% of whom are college-ready. The probability of randomly selecting a college-ready student is .1, but this does not make the probability of Isaac being college-ready equal to .1.

A Particular Numerical Criticism

To turn to Achinstein's example, we are to consider an urn of bags (or populations) such that for each such bag a hypothesis H either is or is not true of it. In particular, if the bag selected is one with 60% white balls, then the hypothesis H_1 is true of that bag. In other words, H_1 plays the role of a one-place predicate $H_1\_\_$, and any particular bag, say b, either has H_1 or not (i.e., either $H_1 b$ or $\sim H_1 b$ is true). Achinstein's criticism assumes:

(*) If p% of the bags have property H_1, then for any randomly selected bag, b, $P(H_1) = p$, (i.e., $P(H_1 b) = p$).

In particular, the bag of balls from which we drew our sample of 100, bag b, "itself is chosen at random from a very large set of bags" (Achinstein, 2001: 134), out of which 99,999 out of 100,000 of the bags have 60% white balls. He infers:

(*) $P(H_1) = .99999$, i.e., P(bag b has 60% white balls) = .9999.

Then applying Bayes's theorem, we get

(1) $P(H_{I}/\bar{X}_{obs} = .4) = .996,$

and he may wish to say

(2) $P(\text{not-}H_{I}/\bar{X}_{obs} = .4) = .004.$

But I have allowed earlier that

(3) $\bar{X}_{obs} = .4$ is evidence that H': $p < .6$.

Therefore, Achinstein's critique continues, Mayo claims we have good evidence that $p < .6$, and so, good evidence that not-H_{I}, even though the posterior probability of not-H_{I} is very low.

Now this is problematic only if two additional premises are true:

> A necessary condition for x to be evidence for H is that $P(H/x) =$ high.

> The prior probability assignment in (*) (upon which the posterior is based) is valid.

I would deny both of these premises. Whereas the first is an issue of philosophy of evidence, the second is based on the fallacy I call *the fallacy of probabilistic instantiation*.

The Fallacy of Probabilistic Instantiation

Since the fallacy is committed repeatedly in mounting these criticisms, let us draw it out a bit. It is just the sort of misstep that one would hope philosophers of probability and statistics would use their acumen to expose, rather than commit.

The basis for the prior probability assignment in (*) is this:

1. Hypothesis H is true of p% of the populations (bags) in this urn of populations U.

2. $P(H$ is true of a randomly selected bag from an urn of bags U$) = p$.

3. The randomly selected bag that was drawn in test T_{I} is b_{I}.

Therefore

(*) $P(H$ is true of $b_1) = p.$

But either H is true of b_1 or not—the probability in (*) is fallacious and results from an unsound instantiation. It may help to make the point using confidence intervals.

An Analogous Fallacy with Confidence Intervals

A 95% confidence interval estimation procedure has a probability of covering the true but unknown value of a parameter equal to .95, in a series of experiments on the same or different populations. Each bag in the pool of bags is a different population, and a 95% confidence interval estimate may be formed for each; however, each interval estimate either will or will not be true of that population. Nevertheless, the foregoing reasoning would countenance the fallacious inference:

1. P (the 95% confidence interval procedure yields an interval estimate that is true) =.95

2. The 95% confidence interval procedure yields an interval estimate: $(.3 < p < .5)$, let us suppose.

Therefore,

(*) $P(.3 < p < .5) = .95.$

But either $(.3 < p < .5)$ or not!

Students from the Wrong Side of Town

Examples of balls in bags, however dear to philosophers, are rather distant from the kinds of realistic examples about which one's intuitions are clearest. It will be useful to turn to some more realistic examples on which the identical criticism has been based, both against frequentist statistics in general and my severity account in particular. Once again, the fallacy of instantiating probabilities is committed.

Our student, Isaac, has passed comprehensive tests of mastery of high school subjects regarded as indicating college readiness. Because such high scores x could rarely result among high school students who are not sufficiently prepared to be deemed college-ready, we regarded x as good evidence for

> H: Isaac is not deficient but is college-ready.

Thus x is good evidence against

> H': Isaac's mastery of high school subjects is deficient, that is, he is not college-ready.

Although as with the binomial example, we would ordinarily consider degrees of readiness, we can keep to this oversimplified rendition to go along with our critic as much as possible.

In this variation of the Bayesian example (from Howson, 1997 and others), we are given that

> $P(x/H$: Isaac is college-ready) is practically 1

whereas

> $P(x/H'$: not college-ready (i.e., deficient)) = very low, say .05.

Given the assumptions that go into the probability calculations are met, it would seem that H has stood up to a fairly severe test.

"But wait a minute!" says the critic. Isaac was randomly selected from a population (perhaps a certain section of the Bronx) wherein college readiness is exceedingly rare—say, only .1% would be correctly described as ready. Accordingly, it is reasoned that

> (*) $P(H) = .001$.

Thus, the posterior probability for H is still low, and for the alternative, H' (deficient), the posterior is high. In particular, to give one illustration, we can have:

> $P(H'/x) = .95$.

Although $P(H/x)$ has increased from $P(H)$, the posterior for H is still low because of the extremely small prior for H.

Notice that if Isaac had been selected from a population where college readiness is rather common—if, say, he was selected from students in a certain affluent neighborhood—then the very same set of passing scores x would now be regarded as strong evidence for H, Isaac being ready. Using this way of evaluating evidence, a high school student from a non-affluent neighborhood would need to have scored quite a bit higher on these tests than one selected from the affluent neighborhood in order for his scores to be considered evidence for his readiness! (Talk about reverse discrimination!)

Once again the same fallacy is committed in arriving at (*), only here perhaps our intuitions that something is amiss are more pronounced.

Although the probability of a random sample (i.e., a student) taken from the "urn" of highschoolers in the given area of the Bronx is .001, it does not follow that Isaac, the one we happened to select, has a probability of .001 of being college-ready (see Mayo, 1997).

CONCLUDING THE ACHINSTEIN DISCUSSION

The foregoing remarks do not preclude saying that the probability of a student randomly selected (from a given population U where p% are ready) is ready equals p—so long as one is clear about the kind of trial is described. What they preclude is treating the probability of the occurrence of this *event* as if it provided a legitimate frequentist probability for the *hypothesis H* in the testing problem. The cavalier manner in which philosophers of probability use the term "hypothesis" makes it too easy to slip between events and statistical hypotheses. A statistical hypothesis, for a frequentist statistician, must describe enough about the data generation G to assign probabilities to all possible outcomes; an event does not.

Nor need we preclude the possibility of a statistical hypothesis *H* having a legitimate frequentist prior. For example, a frequentist probability that Isaac is college-ready might refer to genetic and environmental factors that determine the chance that a high school student (from a specified population) is deficient—something we can scarcely even cash out, much less compute. I return to these points in my concluding comments.

In fairness to Achinstein, it must be noted that he too, presumably, denies that the high posterior for *H* entails that x provides evidence for *H*, but not because he questions the prior probability assignment in (*). He accepts (*) but denies that high posterior probability suffices for evidence. Why then, one might ask, should it even be a necessary requirement; that is, why suppose the computation based on (*) is necessary for evidence for *H*? By assuming a high posterior is necessary, Achinstein must look elsewhere to avoid taking x as evidence for *H* in examples such as test T_1; and he does so by appealing to intuitions that tell him that the right sort of explanatory connection is absent in these examples. By leaving this at a vague intuitive level, however, his account supplies no directions for determining if one really has evidence. We are left to fall back on the very intuitions we wish an account of evidence to provide. Error probabilities, within a severity assessment, let us go further.

For instance, the severity criterion provides the basis for denying that the high posterior $P(H'/x)$ is evidence for H', in the case of Isaac's readiness (see "Students from the Wrong Side of Town," above). The error probability associated with such a procedure for interpreting data would be extremely high. Thus it would lead to regarding data as evidence on the basis of a procedure that very probably would be wrong—low severity! A severity assessment captures and guides intuitions about evidence.

What We *Really,* Really Want Is . . . High Severity!

It is (or should be) well known that error probabilistic concepts, such as p-values and type I and type II errors, do not supply probabilities to statistical hypotheses, and that interpreting them as if they did leads to fallacies, paradoxes, and contradictions. Error probabilities and any severity assessment that we would base upon them, are—quite deliberately—defined exclusively in terms of the *sampling distribution* of $d(X)$, under one or another statistical hypothesis of interest. In contrast, posterior probabilities of a hypothesis such as H_0, conditional on the observed $d(x)$ require a prior probability assignment to (an exhaustive set of) hypotheses. (The capital X indicates the random variable; the lower case x, the resulting value or outcome.) Nevertheless, critics—especially from Bayesian quarters—have long insisted that "what we really want" from tests are posterior probabilities of hypotheses, and some even argue that testers cannot help but fallaciously interpret p-values as supplying a posterior to the null. Those criticisms are analogous to those in "Achinstein's Criticism . . ." (above) and commit analogous fallacies.

A Common Variant on the Criticisms: P-Values versus Posterior Probabilities

The most telling criticisms are put in terms of p-values: Critics argue that (a) certain choices of prior probabilities for the null and alternative hypotheses show that a small p-value is consistent with a much higher posterior probability in a null hypothesis, from which they conclude that (b) significance test reasoning is invalid, or at least is incapable of being used to assess the evidence against the null hypothesis. The criticism assumes that the Bayesian posterior gives the correct or even an acceptable measure of the degree of evidence, reliability, or beliefworthiness properly accorded the null, and thus a conflict between Bayesian and frequentist assessments shows the latter to be at fault! Nowhere is this assumption defended—one is to have a gut feeling that the only way to use data to bear upon the truth of hypotheses is by means of a posterior probability assignment. If the Bayesian posteriors really did provide assessments of the reliability or beliefworthiness of hypotheses, that would be one thing—but they do not.

Achinstein at least is aware of the need to provide grounds for requiring, as necessary for x to be evidence for H, that $P(H/x)$ be high. His reasons are persuasive—*provided that "high probability" equates to "highly warranted in believing."* The trouble is, the prior and posterior probabilities he actually calculates do not warrant this equation. The same will be true for the criticisms from more formal quarters.

(Two-sided) Test of a Mean of a Normal Distribution, Test T_2

The most influential attempts to demonstrate the conflict between p-values and Bayesian posteriors consider a two-sided Normal distribution test, test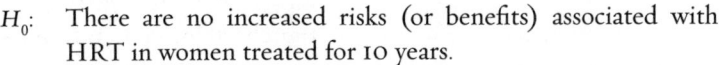
T_2, of H_0: $\mu = 0$ versus H_1: $\mu \neq 0$ (the difference between p-values and posteriors being less marked with one-sided tests), as in Pratt (1965), Berger and Sellke (1987), and Berger (2003). A random sample $X = (X_1, \ldots, X_n)$ is taken where each X_i is distributed Normally with unknown mean μ, and known standard deviation σ and we are testing for discrepancies from 0 in both the positive and negative directions (see Mayo, 2003). We can imagine that null hypothesis H_0 is (the formal embodiment of) the claim:

H_0:   There are no increased risks (or benefits) associated with HRT in women treated for 10 years.

A familiar criticism even with sample size only 50 is: "If $n = 50$ one can classically 'reject H_0 at significance level p = .05,' although $P(H_0/x) = .52$ (which would actually indicate that the evidence favors H_0)" (Berger and Sellke, 1987: 113; we replace Pr with P for consistency) assuming a prior of $P(H_0) = .5$. Note that we would take the low significance level as evidence against H_0 and for H_1, evidence for the very weak claim that there is *some* non-zero difference in risk rates between treated and non-treated. This is taken as a criticism of p-values, only because it is assumed that the .51 posterior in H_0 is the evidence for H_0. (The severity interpretation in the case of a rejection automatically goes on to consider the extent of the discrepancy indicated, but we can put that aside for the present discussion.)

We can concede the critic's point that data we would regard as evidence against H_0 would, on the Bayesian construal, "actually indicate that the evidence favors H_0," at least assuming the recommended prior of .5 to H_0. As the sample size increases, the conflict becomes more noteworthy. If $n = 1000$, a result statistically significant at the .05 level leads to a posterior to the null of .82! It has again gone up, but even more dramatically. Nevertheless, far from discrediting the frequentist assessment, this fact seems to us to count against regarding the Bayesian posterior as properly measuring evidence. This leads to the question, What warrants the prior probability on which the example is based?

The Case of Subjective Priors

Many Bayesians construe prior probability assignments as quantifying their degrees of belief in hypotheses, extracted either by intuition or by strategies of eliciting betting behavior. A strong prior degree of belief in H_0 suffices to ensure that the posterior of H_0 is still high, even with a statistically significant result. By setting the prior high enough, not surprisingly, the o-risk H_0 is saved from counterevidence. That is hardly to subject H_0 to a severe test. Under the pretense of giving a "fair" assignment of priors, assigning .5 to H_0 gives so much weight to H_0 that even highly significant observed departures are taken as strengthening the degree of belief in the null. If understood as a report of an agent's actual degrees of belief, then it is hard to fault, but if we want to know how strongly we *ought* to believe in a hypothesis on the basis of data, the subjective Bayesian analysis will not do.

120

The Assumption of "Objective" Bayesian Priors

Bayesians claim to have arrived at the chart in Table 6.1 and several related charts by appealing to one or another alleged "objective" priors. In particular, following Jeffreys (1939), a prior probability assignment of .5 is given to H_0 and the remaining .5 probability is spread out over the alternative parameter space. Having picked out the point null as a value of special interest to test, their reasoning goes, a "spiked concentration of belief in the null" is "impartial"—but is it? The frequentist says no. Moreover, the "spiked concentration of belief in the null" is at odds with the role played by null hypotheses in testing, where it is so often assumed that "all nulls are false," and we wish only to learn (via significance tests) "how false" the null is. Thus this spiked concentration of belief should hardly be taken as the guidepost from which to launch a critique of significance tests.

Table 6.1 $P(H_0/x)$ *for Jeffreys-Type Prior*

		n						
p	t	1	5	10	20	50	100	1,000
.10	1.645	.42	.44	.47	.56	.65	.72	.89
.05	1.960	.35	.33	.37	.42	.52	.60	.82
.01	2.576	.21	.13	.14	.16	.22	.27	.53
.001	3.291	.086	.026	.024	.026	.034	.045	.124

Source: Berger and Sellke (1987: 113). Used by permission.

Even a discrepancy of several standard deviations away from 0, as Table 6.1 shows, will have little or no chance of being considered as *some* evidence against the null because the posterior probability is below the prior! The error statistician would criticize such tests as having extremely low probability of detecting a false null, that is, low power. Such tests would commit type II errors very often if not in extreme cases with probability 1.[16] This is a remarkable procedure. In the HRT study, there were more than 16,000 women, so the discrepancy between the p-value and posteriors would be even more marked than with n = 1000.

One can see why the Bayesian significance tester wishes to start with a fairly high prior to the null—else, a rejection of the null would be merely to claim that a fairly improbable hypothesis has become more improbable (Berger and Sellke, 1987: 115). By contrast, it *is* informative for an error statistical tester to reject a null, even assuming it is not precisely true, because we can learn "how false" it is. It is all well and good to exhort that what we really want is an assessment of the truth of H_0 given the data— but the Bayesian posterior hardly seems appropriate for the task. A dialogue between a researcher and a Bayesian might proceed as follows:

121

> RESEARCHER: I have found a difference that is significant at the .01 level. Since this would occur only 1% of the time, if in fact there was no effect, it seems to indicate evidence of a genuine effect.
>
> BAYESIAN: This just goes to show what's wrong with significance test reasoning. For a Bayesian who assigned a "fair" prior to the null, one's degree of belief in H_0 would go from .5 to .82.
>
> RESEARCHER: But the null hypothesis would be outside of the 95% confidence interval—surely that is evidence against it, and yet you assign it a probability of .82.
>
> BAYESIAN: Only by calculating a Bayes factor (or related conditional measure) can one judge how well the data supports a hypothesis (Berger and Delempady). The Bayesian analysis tells you that the posterior probability of H_0 is .82—and isn't that what you really want to know?
>
> RESEARCHER: Hmm—I don't think so . . .

"Frequentist" Priors Which Are Not Kosher for Frequentist Error Statisticians

A common gambit by Bayesian critics is to construct examples where allegedly only frequentist priors are used. Again considering a Normal distribution test of $H_0: \mu = 0$ versus $H_1: \mu \neq 0$, Bayesians assure us that even one who insists on a frequentist interpretation of probabilities can see that low p-values may be overstating the evidence against the null. In particular, to construe a hypothesis as a random variable, it is imagined that we sample randomly from a population of hypotheses, some proportion of which are assumed to be true. This is a variation on the technique for erecting a frequentist prior that we saw in "Achinstein's Criticism . . . ," and the same fallacy of instantiating probabilities is committed.

We are to consider a pool of null hypotheses from which H_0 may be seen to belong, and compute the proportion of these that have been found to be true in the past. This serves as the prior probability for H_0. We are then to imagine repeating the current significance test over all of the hypotheses in the pool we have chosen, and the posterior probability of H_0 (conditional on the observed result) will tell us whether the original assessment is misleading. But which pool of hypotheses should we use? Shall we look at all those asserting no increased risk or benefit of any sort? Or no increased risk of specific diseases, such as clotting disorders or breast cancer? In men and women? Or in women only? With hormonal drugs, or any treatments? The percentages "initially true" will vary considerably. Moreover, it is hard to see that we would ever know the proportion of true nulls, rather than merely the proportion that have thus far not been rejected by other statistical tests! (See Mayo, 2003.)

Innocence by Association

Further, even if we agreed that there was a 50% chance of randomly selecting a true null hypothesis from a given pool of nulls, that would still not give the error statistician a frequentist prior probability of the truth of *this* hypothesis, for example, that HRT has no effect on breast cancer risks. Either HRT alters cancer risk or it doesn't. The relevant parameter—say, the increased risk of breast cancer—could conceivably be modeled as a random variable, but its distribution would not be given by computing the rates of other apparently benign or useless treatments! This "frequentist" Bayesian analysis assumes a kind of "innocence by association," wherein a given H_0 gets the benefit of having been drawn from a pool of true or not-yet-rejected nulls. Perhaps the tests have been insufficiently sensitive to detect risks of interest. Why should that be grounds for denying there is evidence of a genuine risk with respect to a treatment (e.g., HRT) that *does* show statistically significant risks?

What Is Varying and What Is Held Fixed

For the error statistician, it must be remembered, what varies is $d(x)$—what is fixed is the particular hypothesis of interest. By contrast, the

Bayesian analysis is conditional on a value of x (other values that could have resulted but did not are irrelevant once x is in hand);[17] X is fixed and the hypotheses are taken to vary.[18] We need not deny that there are contexts wherein that kind of calculation is relevant, but no case has been made that evaluating the truth or correctness of *this* hypothesis is one of those contexts.

Fisher: The Function of the P-Value Is Not Capable of Finding Expression

Faced with conflicts between error probabilities and Bayesian posterior probabilities, the error probabilist would conclude that the flaw lies with the latter measure. This is precisely what Fisher argued, and it seems fitting to finish our story by considering him.

123

Discussing a test of the hypothesis that the stars are distributed at random, Fisher takes the low p-value (about 1 in 33,000) to "exclude at a high level of significance any theory involving a random distribution" (Fisher, 1956: 42). Even if one were to imagine that H_0 had an extremely high prior probability, Fisher continues—never mind "what such a statement of probability a priori could possibly mean"—the resulting high posteriori probability to H_0 would only show, he thinks, that "reluctance to accept a hypothesis strongly contradicted by a test of significance" (ibid: 44) . . . "is not capable of finding expression in any calculation of probability a posteriori" (ibid: 43). Note too that frequentists do not deny there is ever a legitimate frequentist prior probability distribution for a statistical hypothesis. One may consider hypotheses about such distributions and subject them to probative tests. Indeed, if one were to consider the claim about the a priori probability to be itself a hypothesis, Fisher suggests, it would be rejected by the data!

CONCLUDING COMMENTS

In this chapter I have attempted to sketch a conception of evidence as the result of surviving a severe or risky test:

> Data x in test T provide good evidence for inferring H (just) to the extent that hypothesis H has passed a severe test T with x.

On this conception of evidence, even claims that are highly beliefworthy, or highly probable (however probability is construed), would not get credit from tests that poorly probed them. Evidential credit goes only to those specific hypotheses or models that have survived probative testing. I showed how this notion should be cashed out, and contrasted it to other severity accounts, for example, the comparative account of Popper. I then confronted central criticisms raised by philosophers such as Achinstein and Howson, as well as by statistical practitioners such as Berger, Cohen, or Meehl, regarding the error statistical account of tests. Whether the criticisms are directly aimed at severity or, as in statistics, at error probabilistic tests, the criticisms, if sound, would show the inadequacy of the severity account of evidence that I favor. I grant that hypotheses that pass severe tests on my account may be accorded low posterior degrees of support, probability, or belief, while denying this shows any inadequacy in my account.

I have argued, with respect to each variant of the criticism, (1) that the probabilistic assignment commits a fallacy, which may be called the *fallacy of instantiating probabilities*; and (2) that the error statistical assessment of the data, but not the assessment advocated by the critic, is in sync with the goals of severity, and with our intuitions about when data should count as supporting evidence in science. *Highly probed* differs from *highly probable* (in any of the interpretations put forward for the latter), *and it is the former that matters for evidence.*

NOTES

1. Robert A. Wilson, M.D., *Feminine Forever* (M. Evans and Company Inc., New York, 1966).

2. AARP, American Association of Retired Persons, Nov/Dec 2002, p. 72.

3. Minimally $P(x;H) > P(x;\text{not-}H)$. The advantage of this definition is that any measure of evidential relationship, degree of confirmation, probability, etc., can be regarded as supplying a fit measure. Severity can then be assessed by computing the error probability required in (ii).

4. Ben-Haim makes this notion rigorous by means of a definition based on convex sets, but it is one that I do not understand sufficiently to explicate.

5. For example, searching for factors that show statistically significant correlations increases the overall error rate, in contrast to accounts that endorse the likelihood principle. See Mayo and Kruse, 2001; Cox and Hinkley, 1974.

6. This contrasts with the use of type I and type II error probabilities in Neyman-Pearson tests: low type I error is desirable when the null hypothesis is rejected; low type II error, when it is accepted.

7. Of course the severity analysis does not in itself specify the type or size of discrepancies that are of substantive importance, nor what should be done when discrepancies are found. However, this portion of the analysis sets the stage for subsequent strategies for respecifying models to arrive at empirically adequate statistical models (Mayo and Spanos, 2004).

8. This is to be cashed out as "the probability of x under the assumption that H incorrectly describes the procedure that actually generated data x."

9. There are qualifications here that I am omitting.

10. The idea that "same likelihood" means "same evidence" is the thrust of the likelihood principle that underlies "Likelihoodist" and Bayesian accounts (for a full discussion, see Mayo and Kruse, 2001; Mayo, 2002b).

11. Had Popper been aware of the developments in statistics going on right around him in the 1930s and 40s, one suspects the history of the philosophy of science would have been very different. In a letter from Popper in the early 1990s, he expressed regret at having never fully studied statistical methodology.

12. How this contrasts with their traditional construal as mechanical tools for controlling error-rates in the long run is a complex issue that is discussed at length elsewhere (see Mayo, 1996).

13. It must be calculated under one of the two hypotheses. Under H_0 it is .05; under H_1, slightly less.

14. That is the purpose of the "rule of acceptance" in the severity interpretation of statistical tests (no evidence against is not evidence for). See Mayo, 1996.

15. See Mayo and Spanos, 2000.

16. For example, consider a test with significance level .05. Such a test finds evidence for H_0 if the result does not reject H_0, and still finds evidence for H_0 when the usual test rejects H_0 at the .05 level-hence, it has no chance of finding evidence against H_0. Such a test would have 0 severity if used to infer H_0.

17. Such irrelevance of the sample space follows from the Likelihood Principle.

125

18. Admittedly, some frequentist philosophers (Salmon, following Reichenbach) pursued the idea of giving frequentist priors this way; though Reichenbach regarded the ability to make "claims about how often hypotheses like this one are true" as at most a possible situation we might find ourselves in once we had learned enough. Even if we had such numbers, their relevance to reasoning about the hypothesis in hand would be dubious (Mayo, 1996, ch. 4).

REFERENCES

Achinstein, P. (2001) *The Book of Evidence*, Oxford University Press, New York.

Ben-Haim, Y. (2001) *Information-Gap Decision Theory: Decisions Under Severe Uncertainty*, Academic Press, San Diego.

Berger, J. (2003) "Could Fisher, Jeffreys, and Neyman Have Agreed?" *Statistical Science* 18, 2003: 1–12.

Berger, J. O., and T. Sellke (1987) "Testing a Point Null Hypothesis: The Irreconcilability of P Values and Evidence," *Journal of the American Statistical Association*, 82: 112–22.

Birnbaum, A. (1962) "On the Foundations of Statistical Inference" (with discussion), *Journal of the American Statistical Association*, 57: 269–326.

Carnap, R. (1962) *Logical Foundations of Probability*, University of Chicago Press, Chicago.

Casella, G., and R. L. Berger (1987) "Reconciling Bayesian and Frequentist Evidence in the One-Sided Testing Problem," *Journal of the American Statistical Association*, 82: 106–11.

Chalmers, A. F. (1999) *What Is This Thing Called Science?* 3rd ed., University of Queensland Press, Australia.

Cohen, J. (1994) "The Earth Is Round (p < .05)," *American Psychologist* 49 (12): 997–1003.

Cox, D. R. (1981) "Statistical Significance Tests," *British Journal of Clinical Pharmacology*, 14: 325–31.

Cox, D. R., and D. V. Hinkley (1974) *Theoretical Statistics*, Chapman and Hall, London.

Dorling, J. (1979) "Bayesian Personalism, the Methodology of Scientific Research Programmes, and Duhem's Problem," *Studies in History and Philosophy of Science*, 10: 177–87.

Earman, J. (1992) *Bayes or Bust? A Critical Examination of Bayesian Confirmation Theory*, MIT Press, Cambridge, MA.

Edwards, W., H. Lindman, and L. Savage (1963) "Bayesian Statistical Inference for Psychological Research," *Psychological Review*, 70: 193–242.

Efron, B. (1986) "Why Isn't Everyone a Bayesian?" *The American Statistician*, 40:1–4.

Fisher, R. A. (1930) "Inverse Probability," *Proceedings of the Cambridge Philosophical Society*, 26: 528–35.

Fisher, R. A. (1955) "Statistical Methods and Scientific Induction," *Journal of the Royal Statistical Society*, B, 17: 69–78.

Fisher, R. A. (1956) *Statistical Methods and Scientific Inference*, Oliver and Boyd, Edinburgh.

Gibbons, J. D. and J. W. Pratt (1975) "P-values: Interpretation and Methodology," *The American Statistician*, 29: 20–25.

Hacking, I. (1965) *Logic of Statistical Inference*, Cambridge (CVP).

Hacking, I. (1980) "The Theory of Probable Inference: Neyman, Peirce and Braithwaite," pp. 141–60 in D. H. Mellor, ed., *Science, Belief and Behavior: Essays in Honour of R.B. Braithwaite*, Cambridge University Press, Cambridge.

Harper, W., and C. A. Hooker, eds. (1976) *Foundations of Probability Theory, Statistical Inference, and Statistical Theories of Science*, Vol. II, D. Reidel, Dordrecht.

Howson, C. (1997) *Philosophy of Science*, 64: 268–90.

Howson, C. and P. Urbach (1989) *Scientific Reasoning: The Bayesian Approach*, Open Court. La Salle, IL (Second Edition, 1993).

Kyburg, H. E., Jr. (1993) "The Scope of Bayesian Reasoning," pp. 139–52 in D. Hull, M. Forbes, and K. Okruhlik, eds. *PSA 1992*, Vol. II, Philosophy of Science Association, East Lansing, MI.

Kyburg, H. E., Jr., and M. Thalos, eds. (2002) *Probability Is the Very Guide of Life*, Open Court, Oxford.

Laudan, L. (1997) "How About Bust? Factoring Explanatory Power Back into Theory Evaluation," *Philosophy of Science* 64: 303–16.

Lehmann, E. L. (1990) "Model Specification: the views of Fisher and Neyman, and later developments," *Statistical Science*, 5: 160–68.

Lehmann, E. L. (1993) "The Fisher and Neyman-Pearson Theories of Testing Hypotheses: One Theory or Two?" *Journal of the American Statistical Association*, 88: 1242–49.

Lindley, D. V. (1957) "A Statistical Paradox," *Biometrika*, 44: 187–92.

Lindley, D. V. (1976) "Bayesian Statistics," in W. L. Harper and C. A. Hooker, eds. (1976), pp. 353–62.

Mayo, D. G. (1983) "An Objective Theory of Statistical Testing," *Synthese*, 57: 297–340.

Mayo, D. G. (1985) "Behavioristic, Evidentialist, and Learning Models of Statistical Testing," *Philosophy of Science*, 52: 493–516.

Mayo, D. G. (1996) *Error and the Growth of Experimental Knowledge*, The University of Chicago Press, Chicago.

Mayo, D. G. (1997) "Duhem's Problem, The Bayesian Way, and Error Statistics, or 'What's Belief Got to Do with It'?" and "Response to Howson and Laudan," *Philosophy of Science*, 64: 222–24 and 323–33.

Mayo, D. G. (2002a) "Theory Testing, Statistical Methodology, and the Growth of Experimental Knowledge," pp. 171-90 in *Proceedings of the International Congress for Logic, Methodology, and Philosophy of Science*, Kluwer Press, Netherlands.

Mayo, D. G. (2002b) "An Error-Statistical Philosophy of Evidence" and "Response to Professors McCoy and Casella," pp. 79–97, 101–115, in *Scientific Evidence*, Contributions to the Ecological Society of America Conference, University of Chicago Press.

Mayo, D. G. (2002c) "Severe Testing as a Guide for Inductive Learning," pp. 89–117 in H. E. Kyburg and M. Thalos, eds., *Probability Is the Very Guide of Life* (2002), Open Court, Chicago.

Mayo, D. G. (2003) "Could Fisher, Jeffreys, and Neyman Have Agreed? Commentary on J. Berger's Fisher Address," *Statistical Science* 18: 19–24.

Mayo, D. G. (2004) *The Philosophy of Statistics*, Routledge Encyclopedia.

Mayo, D. G., and A. Spanos (2000) "A Post-data Interpretation of Neyman-Pearson Methods Based on a Conception of Severe Testing," *Measurements in Physics and Economics Discussion Paper Series*, History and Methodology of Economics Group, The London School of Economics and Political Science, Tymes Court, London.

Mayo, D. G., and A. Spanos (2004) "Methodology in Practice: Statistical Misspecification Practice," *Philosophy of Science* 71 (5).

Mayo, D. G., and M. Kruse (2001) "Principles of Inference and their Consequences," pp. 381–403 in D. Cornfield and J. Williamson, eds., *Foundations of Bayesianism*, Kluwer Academic Publishers, Netherlands.

Meehl, P. E. (1967/1970) "Theory-Testing in Psychology and Physics: A Methodological Paradox," In D. E. Morrison and R. E. Henkel, eds., *The Significance Test Controversy* (1970), Aldine, Chicago.

Morrison, D., and R. Henkel, eds. (1970) *The Significance Test Controversy*, Aldine, Chicago.

NAS Report (Columbia Accident Investigation Board Report, August 2003, www.caib.us/news/report/default.html).

Neyman, J. (1935) "On the Problem of Confidence Intervals," *The Annals of Mathematical Statistics*, 6: 111–16.

Neyman, J. (1941) "Fiducial Argument and the Theory of Confidence Intervals," *Biometrika* 32: 128–50.

Neyman, J. (1952) *Lectures and Conferences on Mathematical Statistics and Probability*, 2nd ed. U.S. Department of Agriculture, Washington, D.C.

Neyman, J. (1955) "The Problem of Inductive Inference," *Communications on Pure and Applied Mathematics*: 13–46.

Neyman, J. (1957a) "Inductive Behavior as a Basic Concept of Philosophy of Science," 25: 7–22.

Neyman, J. (1976) "Tests of Statistical Hypotheses and their use in Studies of Natural Phenomena," *Communications in Statistics—Theory and Methods*, 5: 737–51.

Neyman, J. (1977) "Frequentist Probability and Frequentist Statistics," *Synthese*, 36: 97–131.

Neyman, J., and E. S. Pearson (1967) *Joint Statistical Papers*, Berkeley: University of California Press.

Pearson, E. S. (1955) "Statistical Concepts in Their Relation to Reality," *Journal of the Royal Statistical Society*, B, 17: 204–07.

Peirce, C. S. (1931–35) *Collected Papers*, Vols. I-VI, ed. C. Hartshorne and P. Weiss; Vols. VII-VIII, ed. A. Burks, Harvard University Press, Cambridge, MA.

Popper, K. (1959) *The Logic of Scientific Discovery*, Basic Books, New York.

Popper, K. (1983) *Realism and the Aim of Science*, Rowman and Littlefield, Totawa, NJ.

Pratt, J. (1965) "Bayesian Interpretation of Standard Inference Statements" (with discussion), *Journal of the Royal Statistical Society*, B, 27: 169–203.

Rosenthal, R. (1994) "Parametric Measures of Effect Sizes," pp. 231–44 in H. M. Cooper and L. V. Hedges, eds., *The Handbook of Research Synthesis*, Sage, Newbury, CA.

Rosenthal, R., and J. Gaito (1963) "The Interpretation of Levels of Significance by Psychological Researchers," *Journal of Psychology*, 64: 725–39.

Royall, R. (1997) *Statistical Evidence: a Likelihood Paradigm*, Chapman and Hall, London.

Salmon, W. (1966) *The Foundations of Scientific Inference*, University of Pittsburgh Press, Pittsburgh.

Savage, L., ed. (1962) *The Foundations of Statistical Inference: A Discussion*, Methuen, London.

7

CONSILIENCE, CONFIRMATION, AND REALISM

Laura J. Snyder

That rules springing from remote and unconnected quarters should thus leap to the same point, can only arise from that *being the point where the truth resides.*[1]
—William Whewell

In the nineteenth century, William Whewell claimed that his confirmation criterion of consilience was a truth-guarantor: we could, he believed, be certain that a consilient theory was true. Since that time Whewell has been much ridiculed for this claim, by critics including J. S. Mill and Bas Van Fraassen. In this chapter I argue that, while Whewell's claim that consilience can *guarantee* the truth of a theory is clearly wrong, consilience is indeed quite useful as a confirmation criterion. Moreover, even when consilience gives evidence for a theory that turns out to be false, there is an important sense in which consilience shows that the false theory has gotten something right. Consilience is a sign that a theory has uncovered something about the natural-kind structure of the physical world. Because of this, I argue that Whewell was correct to claim that consilience provides a "criterion of reality."[2] It does so by providing justification for the claim that we have really "cut nature at its joints." In this way consilience can play a role in an argument for scientific realism.

Natural Kinds and Essences

In his *Philosophy of the Inductive Sciences,* Whewell described consilience in the following way: "The Consilience of Inductions takes place when an Induction, obtained from one *class of facts,* coincides with an Induction, obtained from a *different class.* This consilience is a test of the truth of the Theory in which it occurs."[3] By speaking of "classes" of facts, Whewell implied a connection between his notion of natural classes (or, in our modern terms, natural kinds) and consilience. Whewell did not explicitly discuss this connection, and it has been, for the most part, ignored by other commentators.[4] However, to understand Whewell's test of consilience, and to see its value both as a confirmation criterion and as an argument for realism, we must understand the relation between consilience and natural kinds.

Whewell's belief that there are natural kinds is apparent as early as his 1828 work *An Essay on Mineralogical Classification and Nomenclature,* in which he discussed the controversy between proponents of natural and artificial systems of classifying mineralogical specimens.[5] The mineralogist, Whewell claimed, should seek the "natural and true arrangement of minerals."[6] The grouping of things into classes in science is not merely conventional, on Whewell's view; such groupings are meant to be expressive of a natural ordering in nature. Whewell explained later, in the *Novum Organon Renovatum,* that "there are classifications, not merely arbitrary, founded upon some *assumed* character, but natural, recognized by some *discovered* character."[7] Thus Whewell held that natural classes are discovered empirically, not formed a priori. Scientists, for example botanists, "seek something, not of their own devising and creating;—not anything merely conventional and systematic; but something which they conceive to exist in the relations of the plants themselves;—something which is without the mind, not within;—in nature, not in art."[8] That is, on Whewell's view, scientists seek to "carve nature at its joints," in Plato's famous phrase, not merely to classify things into nominal or conventional groupings.[9]

How are we able to recognize natural kinds? Whewell noted that we are directed to them by resemblances, or shared superficial properties, between individual things. But not every superficial resemblance is relevant to classifying things into natural kinds.[10] This is because the superficial properties are not themselves definitive of natural groupings. In addition to sharing observable properties, certain individuals also share underlying properties, or essences. These essences are taken by Whewell (and modern writers on natural kinds) to be causally responsible for some of the observable properties of individual members of the kind. For example, the superficial properties typically exhibited by members of the kind "water" are these: liquidity at room temperature, colorlessness, odorlessness, taste-lessness. Yet there is also an underlying trait, which we now understand to be the chemical structure H_2O. These superficial properties of water are a result of the chemical structure of water.[11] But other properties of a given sample of water—such as its weight and temperature—are not a result of that chemical structure, and so are not properties shared by all members of the natural kind "water."

Although the shared superficial properties are not used in defining the natural kind term, they are important initially in drawing our attention to the possibility that there may exist a natural kind category of which the instances are members.[12] As Whewell explained, "besides the Natural System, by which we *form* our classes, it is necessary to have an *Artificial System,* by which we *recognize* them."[13] Whewell pointed out that Linnaeus himself recognized that an artificial system, such as his own, was a means to reach a natural system.[14] After we establish that there is a natural kind shar-ing one or more underlying traits, the superficial properties are still useful in helping us recognize possible members of the kind.

What, then, are the "essences" of natural kinds, according to Whewell? An essence of a natural kind is (as Ellis recently described it) "the set of properties or structures in virtue of which a thing is something of the kind it is."[15] Whewell did tell us something about what he believed to be the essential, underlying properties that define certain kinds of natural classes. He claimed in the *History of the Inductive Sciences* that, in the case of plants and other organic beings, the essence is the "general structure and organiza-tion" of the being, especially those organs most important for the preser-vation of life.[16] In the *Philosophy,* Whewell suggested that reproductive relations between individuals are also important.[17] In the case of minerals, it is their chemical composition. Moreover, Whewell explained that "the elementary composition of bodies, since it fixes their essence, must deter-mine their properties. Hence, all mineralogical arrangements . . . must be, in effect, chemical. . . . We may begin with the outside, but it is only in order to reach the inner structure."[18] Thus, Whewell held that the underlying essential trait(s) shared by members of natural kinds are causally responsi-ble for the production of their superficial properties.

In addition to the kinds that are composed of objects or substances, Whewell also spoke of classes of events or processes. For instance, exam-ples of natural classes for Whewell include the classes of planetary motion and optical diffraction. The members of these classes are individual cases of planetary motion (such as the orbit of Mars) and individual instances of optical diffraction. Individual instances of optical diffraction are recog-nized as belonging to this kind by their superficial properties, such as the fact that they involve light appearing to bend around obstacles. But they also share a kind essence. On Whewell's view, then, natural kinds are groupings of individuals—either individual objects or individual instances of events or processes—that are joined together by virtue of sharing a kind essence.

Whewell was rather less explicit on the question of what constitutes the kind essence for classes of events or processes than he was about classes of minerals or animals. Whewell would perhaps suggest that individual cases of optical diffraction share a set of properties which make each case a member of the kind "optical diffraction." These properties include the kind(s) of substance(s) involved in the process, and the kind(s) of changes that occur.[19] These properties, in turn, are linked inextricably to the *cause* of optical diffraction. Thus, in the case of event or process kinds, the kind essence can be seen as being constituted by the cause or causes of that event or process. The kind essence "determines," as Whewell put it, the similarity of observed properties in each case of optical diffraction. The fact that a group of events constitutes a natural event kind can be inferred even before it is known what the essence of the kind is, just as we can be alerted to the existence of an object kind by shared superficial properties, even before we know what underlying trait(s) constitute the essence of the kind. For instance, gold was assumed to be a natural kind, even before its chemical composition was known. Whewell would say (were he being explicit about this) that the class "optical diffraction" was assumed to be a natural kind, and not just a conventional one, even before its essence was known.

CONSILIENCE AND THE UNIFICATION OF EVENT KINDS[20]

As we saw earlier, consilience occurs when "an induction, obtained from one class of facts, coincides with an induction, obtained from a different class." Whewell also characterizes consilience as a "jumping together of inductions." To understand what Whewell means by this, it may be helpful to schematize the "jumping together" that occurred in the case of Newton's law of universal gravitation, Whewell's exemplary case of consilience. On Whewell's view, Newton used the form of inference Whewell characterized as "discoverers' induction" in order to reach his universal gravitation law, the inverse-square law of attraction. Part of this process is portrayed in Book III of the *Principia,* where Newton listed a number of "propositions." These propositions are empirical laws that are inferred from certain "phenomena" (which are described in the preceding section of Book III).

The first such proposition or law is that "the forces by which the circumjovial planets are continually drawn off from rectilinear motions, and retained in their proper orbits, tend to Jupiter's centre; and are inversely as the squares of the distances of the places of those planets from that centre." The result of another, separate induction from the phenomena of "planetary motion" is that "the forces by which the primary planets are continually drawn off from rectilinear motions, and retained in their proper orbits, tend to the sun; and are inversely as the squares of the distances of the places of those planets from the sun's centre." Newton saw that these laws, as well as other results of a number of different inductions, coincided in postulating the existence of an inverse-square attractive force as the cause of various phenomena.[21] According to Whewell, Newton saw that these inductions "leap to the same point;" that is, to the same law.[22] Newton was then able to bring together inductively (to "colligate") these laws, and facts of other kinds (such as the event kind "falling bodies"), into a new, more general law, namely the universal gravitation law: "All bodies attract each other with a force of gravity which is inverse as the squares of the distances." By seeing that an inverse-square attractive force provided a cause for different event kinds—for satellite motion, planetary motion, and falling bodies—Newton was able to perform a more general induction to his universal law. He could then extend this law to a further event kind, tidal activity.[23]

In terms of natural kinds, what Newton found was that these different event kinds—including circumjovial orbits, planetary orbits, and the class of falling bodies—share an essential property, namely, the same cause. Newton discovered that what makes "the orbit of Mars" a member of the class "planetary orbits" is that it is caused to have the properties it does by an inverse-square attractive force of gravity between Mars and the sun. He also found that other event kinds share this kind essence (i.e., an inverse-square attractive force). What Newton did, in effect, was to subsume these individual event kinds into a more general kind comprised of sub-kinds sharing a kind essence, namely being caused by an inverse-square attractive force. Consilience of event or process kinds results in *causal unification*.[24] More specifically, it results in unification of natural kind categories based on a shared cause. In this case, phenomena that constitute different event kinds, such as "planetary motion," "tidal activity," and "falling bodies," were found by Newton to be members of a unified, more general kind: "phenomena caused to occur by an inverse-square attractive force of gravity" (or, "gravitational phenomena").[25]

According to Whewell, consilience is a sign that we have found a "vera causa," or a "true cause," that is, a cause that really exists in nature, and whose effects are members of the same natural kind.[26] Moreover, by finding a cause shared by phenomena in different sub-kinds, we are able to colligate all the facts about these kinds into a more general causal law. Whewell claimed that "when the theory, by the concurrences of two indications . . . has included a new range of phenomena, we have, in fact, a new induction of a more general kind, to which the inductions formerly obtained are subordinate, as particular cases to a general population."[27] He noted that consilience is the means by which we effect the successive generalization that constitutes the advancement of science.[28]

Note that consilience is importantly different from the type of reasoning known as "inference to the best explanation," even when the "best explanation" is defined as a causal hypothesis.[29] The causal law expressing the essence of different event or process kinds is not one postulated merely because it explains or accounts for these different classes of facts. Rather, in the case of each class, the law has emerged from a process of inductive reasoning (which, for Whewell, consists in various kinds of ampliative inference, including causal, analogical, and eliminative inference).[30] Thus the causal law is not imposed from above as a means of tying together different event kinds. Instead, it wells up from beneath in each separate case. What is important, in the case of consilience, is not merely that a single causal mechanism or law can explain or account for different event kinds, but rather that separate lines of induction lead from each event kind to the same causal mechanism or law—that there is a convergence of distinct lines of argument. At each step, there is inductive warrant for the causal law besides its explanatory utility.

In addition to Newton's law of universal gravitation, Whewell sometimes cited the wave theory of light as an example of a consilient theory. However, Whewell was rather ambivalent over the extent to which the wave theory of light was consilient. (This may be somewhat surprising, because Whewell is generally considered one of the wave theory's most keen supporters in the nineteenth century.[31]) For example, in his Bridgewater Treatise of 1833, Whewell noted that for proponents of the wave theory (presumably including himself) "the theory of undulations is conceived to be established in *nearly* the same manner, and *almost as certainly*, as the doctrine of universal gravitation."[32]

Whewell's tentativeness about the wave theory can be seen by comparing his "Inductive Tables" of astronomy and optics. In introducing these tables, Whewell admitted that the table of astronomy shows the convergence toward simplicity (and thus consilience) in a "greater degree" than that of the wave theory of optics.[33] And indeed, in the case of optics, the table is somewhat incomplete. There are numerous entries reading "imperfect colligations," and, while the base of the table for astronomy includes both the name of the theory ("The Theory of Universal Gravitation") and a causal law ("All bodies attract each other with a force of *Gravity* which is inversely as the squares of the distances"), in the case of optics the base

134

includes only the name of the theory ("The Undulatory Theory of Light") and no specific law. Later, in his 1858 *Novum Organon Renovatum,* Whewell noted that "Kepler discovered that the planets describe ellipses, before Newton explained why they select this particular curve, and describe it in a particular manner [i.e., because they act in accordance with the law of universal gravitation]. The laws of reflection, refraction, dispersion, and other properties of light have long been known; the causes of these laws are at present *under discussion.*"[34]

Why, in 1858, was a proponent of the wave theory claiming that the causes of the laws were still "under discussion"? Surely the cause of each of these phenomena, according to a wave theorist, was the propagation of transverse vibrations through the ether. I propose that Whewell was referring to the fact that there was, as yet, no single specific (mathematical) causal law uniting the various event kinds, as there was in the case of Newton's inverse-square law. This is why Whewell was not prepared to claim unambivalently that the wave theory of light was consilient. What Whewell wanted was perhaps something like the equations for the electromagnetic field provided by Maxwell in his Third Memoir of 1865.

135

There was another problem for the wave theory as well, one of which Whewell was well aware. The wave theory still seemed to be (in the complaint Whewell used against Descartes' vortices), "inconsistent mechanically": the ether seemed both highly elastic (in order to carry rapid vibrations at high velocities) and absolutely solid (because only solids were known to transmit transverse waves). Moreover, the wave theorists needed to suppress the longitudinal component of waves transmitted through an elastic solid. Wave theorists still needed to work out the details of the inferred ether and the waves supposedly being carried through it.

We are now in a position to see why a common argument against Whewell's criterion of consilience is erroneous. Many commentators have agreed with Larry Laudan's claim that consilience is fundamentally a relativistic or contextual criterion. According to Laudan, whether some theory is consilient depends upon our knowledge state with respect to the classes involved; specifically, whether we know that certain classes of facts are members of the same general type of thing.[35] Laudan notes that for Descartes, planetary and terrestrial motions were phenomena of the same type, because both were the result of vortices; thus, Laudan argues, relative to the Cartesian system, Newton's theory was *not* consilient.[36] However, as we have seen, the criterion for consilience is that a theory connects two (or more) different natural kinds into a more general natural kind *by virtue of inductively inferring* the same cause for each natural sub-kind. The issue, then, is not the relativistic one of whether or not we previously *knew* or *suspected* that these sub-kinds all belong to a more general kind, but rather whether they do all fit into the same more general kind, by sharing the same kind essence. Moreover, in the case of consilience we must come to the conclusion by the proper sort of inductive reasoning (thus conferring on the conclusion inductive warrant), and not merely by postulating a common cause hypothetically. Whewell famously claimed that Descartes did not reach his conclusion by proper induction. On the other hand, Whewell believed that Newton's theory was consilient because it satisfied the conditions of a proper induction, and because it causally unified different natural kinds under a more general kind, by showing that they shared a kind essence (i.e., the same cause). The relevant consideration is whether a theory results in this causal unification, and does so by the proper kind of reasoning, not whether natural kinds thus unified were ever previously suspected of being united.

136

This erroneous criticism of Whewell's criterion of consilience may arise from the conflation of Whewell's view with that of his friend, John Herschel. Herschel's brief discussion of consilience in his *Preliminary Discourse* focuses on the subjective, psychological aspects of consilience, and does not involve any requirement of causal unification under more general natural kinds. Herschel explained that "The surest and best characteristic of a well-founded induction . . . is when verifications of it spring up, as it were, spontaneously into notice, from quarters where they might be least expected, or even among instances of that very kind which were at first considered hostile to them."[37] Thus Herschel's view of consilience, which concentrates only on the element of surprise, is prone to Laudan's criticism. On the other hand, although Whewell did at times note the psychological impact of consilience, pointing to its "unexpected coincidence of results,"[38] this was not how he characterized the logical structure of the confirmation criterion. Whewell's consilience may *cause* the psychological element of surprise, but it is *defined by* the logical element of causal unification of different event or process kinds into more general kinds, by virtue of sharing a common cause.

CONSILIENCE'S VALUE AS A CONFIRMATION CRITERION

Whewell made two arguments for the value of consilience as a confirmation criterion. (Later, we will see that both of these can be useful in making an argument for natural-kind realism.) First, Whewell claimed that consilience is "a criterion of reality, which has never yet been produced in favour of falsehood." Elsewhere he similarly noted that "there are no instances, in which a doctrine recommended in this manner has afterwards been discovered to be false."[39] He can thus be seen as making an inductive argument in favor of the confirmation value of consilience, namely, the argument that since consilient theories in the past have always turned out to be true, we can infer that (probably) all consilient theories are true, and thus that our current consilient theories are (probably) true.[40] He has been much ridiculed for this inductive argument, for example by Van Fraassen.[41]

As Whewell's detractors like to point out, the one positive instance strongly adduced by Whewell for his inductive argument has been shown to be a counterinstance: Newton's theory, which seemed irrefutable in Whewell's time, has since been shown to be false. Newton's theory employs conceptions that, it turns out, are not true of the world, such as absolute space and absolute time. The cause postulated by the theory, the inverse-square attractive force of gravity propagated through space and time, is also nonexistent (it is not a *vera causa*). Thus, although Newton's theory was highly consilient, more so than any other theory of Whewell's time—it brought together different event kinds, by having inductive warrant for the inference that they shared a common cause—it turned out to be false. Whewell's inductive argument initially seems to fail to establish any confirmatory value for consilience. However, as we will see later in this section, there is an overlooked sense in which Whewell's claim was clearly right, and this sense should be of particular interest to the natural-kind realist.

Whewell also gave a second argument for the value of consilience as a confirmation criterion. Whewell claimed that "no accident could give rise to such an extraordinary coincidence" as is found in cases of consilience. Whewell compared this to the coinciding testimony of two witnesses.[42] The most probable cause for the agreement of their testimony, Whewell argued, is that what they each claim is the truth. This argument for the value of consilience is therefore a type of reasoning known as the "inference to a common cause." In the case of the consilience of Newton's theory, the inference to a common cause has this form:

1. The cause of the elliptical shape of planetary orbits (Kepler's first law) has been inferred from the data to be an inverse-square force of attraction between the sun and the planets.

2. The cause of the inequalities of the moon's motion has been inferred from the data to be an inverse-square attractive force between the earth and the moon.

3. The cause of the motion of the satellites of Jupiter and Saturn has been inferred from the data to be an inverse-square attractive force between the planets and their satellites.

4. The most probable cause for the correlation or "matching" of results in 1–3 is that the phenomena in 1–3 have a common cause: they share a common causal structure.

Hence,

5a. The event kinds listed in 1–3 above are each members of a more general event kind by virtue of sharing the same cause; and

5b. The cause shared by this more general event kind is the inverse-square attractive force operating between members of the kind.

Conclusions 5a and 5b are both solidly supported, given the common cause inference in number 4. That the different event kinds share a cause follows from the fact that each of the individual lines of inference in 1–3 has led to the conclusion that it is probable that the cause in each case is an inverse-square attractive force. That this shared cause is an inverse-square attractive force follows from the fact that it was inferred to be probable in each of the cases. And, trivially, by the characterization of event kinds as groups of individuals whose members share the same essence, which is the same cause for the events, it follows that the event kinds in 1–3 are members of the same more general event kind. Whewell's criterion of consilience can thus be seen to have a virtue absent from the Reichenbach-Salmon formulation of the inference to a common cause: namely, a means for inferring not only that there is a common cause, but also for inferring what that common cause is. For Whewell, there is independent (though not conclusive) reason for thinking that an inverse-square law, rather than something else, is the cause for the convergence of inductive results in each of the cases. The reason is that, in each case, the inferred cause of the members of the event kind was an inverse-square attractive force.[43]

This common cause argument (with the requirement of the inductive method for reaching the specific cause in each case) does show that consilience bestows justification on theories. Theories that are truly consilient are well-confirmed. Thus, consilience did provide good reason for thinking that Newton's theory was true. The fact that it turned out to be false does not diminish the truth that, relative to what was known at the time, Newton's theory was very well-confirmed.

Of course, the fact that Newton's theory turned out to be false does show that Whewell went too far in claiming that consilience is conclusive evidence for a theory, in the sense of proving that a theory will *never* be shown to be false. In some ways, it would have been more consistent with his general view of the progress of science for Whewell to have claimed that hypotheses satisfying the condition of consilience are our best theories, but are still subject to correction. Whewell suggested this approach in his discussion of Newton's Rules of Reasoning, claiming that "the really valuable part of the Fourth Rule is that which implies that a *constant verification,* and, if necessary, rectification, of truths discovered by induction, should go on in the scientific world. Even when the law is, or appears to be, most certainly exact and universal, it should be constantly exhibited to us afresh in the form of experience and observation."[44]

CONSILIENCE AND REALISM

I argue briefly here that Whewell's confirmation criterion of consilience, in addition to being a powerful confirmation criterion, also provides an argument for natural-kind realism. My claim is that a consilient theory, even when false, gives strong reason for believing that it has pointed to really existing natural classes or kinds, whether or not we have grasped correctly what is the essential property shared by members of these kinds. This form of realism is modest, in that it remains agnostic about the ontological status of the theoretical entities postulated by theories, as well as about specifics regarding the kind essences. Yet it is not overly modest; it does not claim that it is *impossible* to gain knowledge of these theoretical entities and essences.[45] My modest realism is not intended to be restrictive, in the sense of claiming that natural-kind structure is all that can be known. It claims only that we have good grounds for believing that we have knowledge of this natural-kind structure, better grounds than for believing in the reality of the theoretical entities postulated by theories. I propose this realism in the spirit of a modest epistemic effort to provide a foothold for further arguments in favor of realism—perhaps even stronger forms of realism.[46]

Consilience provides two ways to argue for this admittedly modest realist position, and hence for realism itself. These two arguments mirror the strategies used by Whewell in arguing for consilience's value for confirmation. The first, inductive, argument was shown to be problematic for his claim that consilience is a truth-guarantor, given that Newton's theory turned out to be false (though it is still a valid argument—albeit weak, because it has only one positive instance—for the claim that consilient theories are well-confirmed). Nevertheless, an inductive strategy can be useful as part of an argument for realism. Although Newton's theory turned out to be false, the theory still seems to have been correct in bringing together event kinds that belong to a more general kind, showing something true about the causal order or structure of nature, even though it was wrong about what the shared kind essence was. That is, Newton showed that the phenomena of free fall, pendulum motion, lunar acceleration, satellite motion, and planetary motion share a cause, and in this way constitute a general event kind. This insight has not been proved false, although the particular shared cause he postulated has been rejected. (In the numbered list presented earlier, although 5b has turned out to be false, 5a still seems to be true.) Einstein's general theory of relativity proposed instead that the cause of these phenomena (and others) is the curved structure of space-time. It may well be that this proposed cause will be replaced by another (quantum gravity?), but it seems unlikely that a later theory will show that these different phenomena are unconnected, and are not members of the same general event kind. These event kinds do seem to be subkinds of a more general kind that are connected in nature. Those who ridicule Whewell for his belief that consilience never has and never could give evidence for a false theory do not give him the credit that is due, by recognizing that the one truly consilient theory touted by Whewell is still considered to have gotten things right, at least in terms of the natural-kind structure of the physical world.

140

Moreover, this inductive argument can help defeat the so-called "pessimistic induction" used by Laudan against the realist position. The most intuitively powerful argument in favor of realism—advanced with great effect by Richard Boyd and others—is the success of our best theories. This "success" has been characterized in various ways—as success in explanation, prediction, or novelty—but the intuition is the same: the success of science would be difficult, if not impossible, to account for, unless mature scientific theories were at least approximately true (as Putnam said, it would be a "miracle.")[47] This intuition was dealt a blow by Laudan with his "pessimistic induction," that is, his assertion that the history of science contradicts this intuition. Laudan claimed that since most of the "successful" theories of the past turned out to be false, our successful theories will probably turn out to be false as well.[48] However, one conclusion we can draw from this discussion is that if success is defined as consilience, then success is not defeated by the pessimistic induction.

Theories that have been fully consilient according to Whewell's notion (such as Newton's theory), have been supplanted by other theories, yet have not been completely overturned. Parts of the theory have been retained, specifically the natural-kind structure postulated by the theory.[49] Although I will not argue this here, Faraday's and Maxwell's consolidation of electricity, magnetism, and optics is another likely case. It is surely no coincidence that Whewell was extremely interested in Faraday's work, corresponding with him and suggesting terminology and even experiments for him to perform, and that Whewell's philosophy was an important influence on Maxwell.[50] After Maxwell, it seems likely that any future theory of electromagnetism will also need to colligate optics. There are other instances as well. Indeed, it seems fair to say that there has been a striking amount of continuity over the history of science at the level of basic structure, in this sense of grouping together natural kinds whose members share the same cause. Whewell himself referred to this kind of continuity when he claimed that the history of science has consisted in successive generalizations and increasing levels of consilience. By using Whewell's notion of consilience as supplying the criterion of success, it is possible to defeat Laudan's pessimistic induction as an argument against the form of realism I have proposed. Thus the strong intuition in favor of realism is not itself defeated, at least in the case of a modest form of natural-kind realism.[51]

This intuition, however, is grounded in a type of reasoning considered suspect by many, especially as a means of arguing for realism, namely, an "inference to the best explanation." That is, the intuition appealed to in favor of realism is that the best explanation for the success of our finest theories is that these theories do match the way the world is, to some degree. Proponents of anti-realist views have claimed that it is invalid to use IBE to argue for realism, since the very validity of this form of reasoning is what is being questioned by the anti-realist (Fine, Laudan, Van Fraassen).[52] Others have claimed that this form of reasoning is used in judging the empirical adequacy of theories on the basis of observed success, and thus even by the empiricist. Accordingly, critics of realism must show why IBE is acceptable in some contexts and not in others (Boyd, Niiniluoto, Psillos).[53] Still others have asserted that there are no viable anti-realist explanations for the success of science, and so the inference to the best explanation is actually an inference to the *only* explanation (Leplin, Niiniluoto).[54] This dispute will not be solved here. Rather, I will now briefly show that consilience provides an additional way to argue in favor of natural-kind realism, one that does not depend on the disputed form of argument.

We saw earlier that Whewell used the form of reasoning known as "common cause inference" to justify the value of consilience as a confirmation technique. It is also possible for the realist to use this type of inference to argue that consilience gives a good reason for thinking that our consilient theories do accurately depict part of the natural-kind structure of the natural world. Roughly, the common cause argument here asserts that the coincidence of results that occurs with consilience (the same causal law inferred from facts about different event kinds) is most probably the result of a shared cause. Since there is no observed cause for this correlation, it must be that there is an unobserved cause. (We might, of course, be wrong; a common cause inference licenses only that it is *reasonable* to infer the existence of a common cause, not that a common cause is *entailed* by the correlation.)[55] Further, we have reason to believe that this unobserved cause is an underlying natural-kind structure of the physical world, a structure that we have gained insight into. Our reason is that each of our inductions has led to the conclusion that it is probable that there is such a structure, and that it has a certain form.

To a certain degree, this strategy is similar to the one Salmon described as an "empirical argument for realism." According to Salmon, Jean Perrin used this approach in arguing for the existence of molecules in 1908. Different experiments, combined with certain background assumptions, yielded the same value of Avogadro's number (6×10^{23}). Roughly speaking, Salmon claimed that Perrin argued that because the same value was reached in different ways, it was highly probable that molecules really exist and were being counted by the different methods.[56] That is, the common cause for the convergence of results was the existence of molecules. Whewell's consilience provides a similar kind of argument. Because different lines of reasoning over different domains have led to the same natural-kind structure of the world, it is probable that the world does have that structure; the real existence of this structure provides the common cause for the convergence of inductive results. It remains possible, certainly, that we are wrong about what the common cause is. Nevertheless, we have reason to believe that we are right, and that we have uncovered some aspect of the natural-kind structure in nature. Thus consilience provides a common cause argument for the type of natural-kind realism I have proposed.

It might seem that this argument for a modest realism is, in the end, too modest. For one thing, it is not a conclusive argument for realism, nor is it an argument for a strong form of realism. Moreover, by basing an argument for realism upon "success" defined in terms of a very strong form of consilience, it might be argued that I am attributing success at uncovering the causal structure of the natural world to too few theories. But why should it be surprising that only a few theories have managed to uncover the causal structure of nature, especially when some argue that it is impossible to do so at all? As an argument against the pessimistic induction, it is useful to point to the cases where it is impossible to deny that there is some important continuity over the history of science, that is, where successor theories have retained an important claim about the causal order of things. Moreover, such cases also function as initial material for a common cause inference to natural-kind realism. Further historical work will, I believe, increase the stock of theories that are seen as being consilient, and thus add to the evidence for both kinds of argument.

Because of this, I suggest that a useful project for the realist is to try to determine whether there are other cases of scientific theories that satisfy Whewell's notion of consilience, and to what degree the natural-kind structure these theories have postulated has been retained in the later theories that have succeeded them.[57] As suggested above, the consolidation of electricity, magnestism, and optics would be a good starting point, and there are surely other cases as well.[58] In the end, the realist might choose modesty more as a virtue than a necessity. Consilience will then be recognized as not only a method for testing scientific theories, but also a way to argue for scientific realism.

ACKNOWLEDGMENTS

Earlier versions of parts of this chapter were improved by comments from audiences at the University of Minnesota and the Center for Philosophy of Science at the University of Pittsburgh, in the Fall of 2002 and the Spring of 2003, respectively. I greatly appreciate the helpful feedback, especially from Michel Janssen, Jim Lennox, Sandra Mitchell, and John Norton. I also thank Peter Achinstein and Giovanni Boniolo for comments on an earlier draft. My work on this chapter was supported by a summer research grant from St. John's University.

143

NOTES

1. *Philosophy of the Inductive Sciences*, II, p. 63.

2. *Philosophy of the Inductive Sciences*, II, p. 68.

3. From the *Novum Organon Renovatum*, the second volume of the third edition of the *Philosophy of the Inductive Sciences*, pp. 87–88.

4. The exceptions to this are Herschel 1841, Ruse 1976, and Harper 1989; however, none of these writers has closely examined Whewell's notion of natural kinds.

5. Hacking credits Whewell with introducing the issue of kinds to nineteenth-century British philosophy in his discussion of natural and artificial systems of biological classification in the 1841 *Philosophy* (see Hacking, 1991, p. 111). In fact, however, Whewell discussed the issue in his 1828 *Essay on Mineralogical Classification*. Whewell did not use the term "natural kinds," as Hacking notes; Venn introduced this term in his 1866 *Logic of Chance*. Whewell used instead the terms "natural classification" and "natural groups of kinds."

6. Whewell, *Essay on Mineralogical Classification*, p. i.

7. Whewell, *Novum Organon Renovatum*, p. 17. See also p. 220.

8. Whewell, *Philosophy of the Inductive Sciences*, I, p. 492. See also the *Novum Organon Renovatum*, p. 230.

9. Whewell's view of natural kinds, like that of most writers today, is derived from Locke's discussion in Book III of the *Essay on Human Understanding*. Locke there defined natural kinds by the sharing of a real essence, in the sense of some underlying structure or trait causally responsible for the superficial properties of a thing (see III, vi, 2, p. 439). Yet Locke (unlike Whewell) believed that real essences are unknowable to us; thus we can never have knowledge of real or natural kinds (see III, vi, 9, pp. 444–45).

10. Whewell, *Philosophy of the Inductive Sciences*, I, pp. 486–87; see also pp. 513–14.

11. Kornblith, for example, makes this point in his 1993, p. 37.

12. As Putnam put it, we recognize things by their nominal essence or the "stereotype" of a term, and not by their real essences or "extension" of a term; this extension is determined by "experts." Descriptions of superficial properties do not determine whether an individual is a member of a natural kind (deep structure does that); they only "fix the reference of a term." That is, they let the hearer know what type of stuff is being referred to (1975, p. 205).

13. Whewell, *History of the Inductive Sciences*, III, p. 204.

14. See Whewell, *History of the Inductive Sciences*, III, pp. 267–68.

15. Ellis 2000, appendix B.

16. Whewell, *History of the Inductive Sciences*, III, p. 281.

17. See Whewell, *Philosophy of the Inductive Sciences*, I, p. 505. Whewell referred to Cuvier's claim that a species can be defined as "the collection of individuals descended from one another, or from common parents," noting that "we consider it a proof of the impropriety of separating two species, if it be shown that they can by any course of propagation, culture, and treatment, the one pass into the other."

18. Whewell, *History of the Inductive Sciences*, III, pp. 196–97. Whewell also noted that the idea of "Likeness cannot lead us to a real arrangement: this notion requires to have precision and aim given it by some other relation;—by the relation of Chemical Composition in minerals, as by the relation of Organic Function in vegetables" (*Philosophy of the Inductive Sciences*, I, p. 513).

144

19. In unpacking Whewell's implicit view of event kinds, I follow (to some degree) Ellis' view of such kinds. I do not go so far as to suggest that Whewell would agree with the dispositional element of Ellis' essentialism, however.

20. I will here concentrate on Whewell's criterion of consilience as applied to the verification of laws regarding events or processes such as "diffraction" and "tidal activity," rather than on the verification of laws regarding objects such as classes of minerals.

21. As McMullin notes, the notion of the "attractive force" was problematic for Newton; he was never satisfied in his search for an "agent cause" of gravitational behavior, a "cause of gravity," as Newton put it. But Newton did succeed in causally unifying disparate phenomena by the concept of this attractive force (see 2001, p. 296). It is this aspect of Newton's theory that Whewell pointed to as the exemplar of consilience.

22. Whewell, *Novum Organon Renovatum*, p. 88.

23. This type of extension of a causal law to a different event kind, without ad hoc alteration, is what Whewell calls "coherence." Due to lack of space I will not discuss coherence here.

24. Interestingly, it has not generally been noted that the unification that occurs among phenomena by consilience is specifically causal unification. For example, Friedman denies that Whewell applies his notion of consilience to cases of causal laws (see 1983, p. 242, n. 14).

25. Writers on natural kinds agree that kinds may be subsumed under more general kinds, as long as there is no partial overlap; indeed, taxonomical systems such as those found in biology and mineralogy are systems of arranging things into non-overlapping kinds and sub-kinds.

26. See Whewell, *Philosophy of Discovery*, p. 191. Ruse agrees that Whewell's notion of the vera causa lies at the center of his criterion of consilience (1976, p. 232). See also Butts, 1977, p. 81; and Ruse, 1975, pp. 161–62.

27. Whewell, *Novum Organon Renovatum*, p. 96.

28. See Whewell, *Philosophy of the Inductive Sciences*, II, p. 74.

29. Numerous commentators on Whewell have interpreted his view as a form of "inference to the best explanation" or as the related view of abductive or retroductive inference, including Peirce (who coined the term "abduction") 1865; Laudan, 1971; Fisch, 1991, p. 168, n. 13; McMullin, 1992; and Janssen, 2002.

30. For more on Whewell's view of induction, see Snyder, 1997; 2004: forthcoming.

31. Cantor, for example, claims that Whewell was committed to the truth of the wave theory as early as 1831 (see 1983, pp. 197–99), while Buchwald dates Whewell's embrace of the theory to 1832 (1989, p. 296).

32. Whewell, *Astronomy and General Physics Considered with Reference to Natural Theology*, p. 105, emphasis added.

33. *Philosophy of the Inductive Sciences*, II, p. 78.

34. Whewell, *Novum Organon Renovatum*, p. 118; emphasis added.

35. See Laudan, 1971; Forster, 1988; Harper, 1989; and Morrison, 1990.

36. Laudan, 1971: p. 374. For similar claims see Butts, 1977, pp. 74–75; Fisch, 1985; and Morrison, 1990.

37. See Herschel, *Preliminary Discourse*, p. 170.

38. Whewell, *Philosophy of the Inductive Sciences*, II, p. 67.

39. Whewell, *Novum Organon Renovatum*, p. 90, and *Philosophy of Discovery*, p. 192.

40. Assuming the historical claim were correct, how could it have any bearing on the confirmation of hypotheses? If it were the case that every past consilient theory has been true, we could make an inductive argument to the generalization "all consilient theories are true." This generalization would be supported by its instances, as are all universal generalizations. The greater the number of positive instances, the closer to one is the probability that the generalization is true—and the higher the probability that the next consilient theory will be true (for a proof of the relevant theorem of the probability calculus, see Earman, 1985, p. 529). The question remains, though, how this type of inductive argument can play a role in confirmation. If we infer, from the generalization "all crows are black," that "the next crow will probably be black," we have not thereby *confirmed* the hypothesis that the next crow will be black. Whewell's inductive argument could function as a plausibility argument helping to set the prior probability of a hypothesis before other empirical testing. Salmon makes a similar claim in favor of certain virtues in terms of his frequency theory of probability, in his 1970. He claims that we can consider the prior probability of a hypothesis as a "frequency—e.g., the frequency with which hypotheses relevantly similar to the one under consideration have enjoyed significant scientific success" (pp. 85–86). Putting his analysis into more general (non-frequentist) inductive terms, and applying it to the case of Whewell's consilience criterion, we could argue that since all or most relevantly similar (i.e., consilient) theories have been true, we can assign a high prior probability to a given consilient theory.

145

41. See Van Fraassen 1985, p. 267.

42. Whewell, *Philosophy of Discovery*, p. 190. See also Whewell, *An Essay on Mineralogical Classification*, p. iv; and Salmon 1984, p. 220.

43. For a similar criticism of the Reichenbach-Salmon view, see Sober 1989, pp. 281–82.

44. Whewell, *Philosophy of Discovery*, p. 196, emphasis added.

45. Thus, my modest realism rejects Poincaré's claim that it is *impossible* to have knowledge of the theoretical entities postulated by scientific theory: "still things themselves are not what [science] can reach as the naive dogmatists think, but only relations between things. Outside of these relations there is no knowable reality" (cited in Psillos, 1999, p. 150).

46. There are, of course, many different varieties of scientific realism. For a recent survey, see Achinstein, 2002.

47. See Boyd, 1984, and Putnam, 1975; 1978. For more recent versions of this argument, see Carrier, 1993 and Kornblith, 1993.

48. See Laudan, 1981; 1984. Mary Hesse had made a similar point earlier in her 1976.

49. Similar strategies for arguing against Laudan have been employed by Worrall, 1989; Kitcher, 1993; and Psillos, 1996. All argue that it is necessary to distinguish between parts of theories that have been abandoned, and parts that have been retained, when the theories have been superseded by newer ones.

50. For the relation between Whewell and Faraday, see Snyder, 2002. For Whewell and Maxwell, see Fisch, 1988 and Harman, 1998.

51. Since I am not here arguing for a strong type of realism in which the theoretical terms of our best theories are taken to refer to really existing unobservable entities, it may be that Laudan would claim that my natural-kind realism is a form of "realism without the real." Obviously I disagree. See Laudan, 1984, a response to Hardin and Rosenberg, 1982.

52. See Fine, 1996; Laudan, 1981, and Van Fraassen, 1980. Even Peter Lipton, a supporter of IBE in science, claims that the use of IBE as an argument for realism is an "abuse" of this form of reasoning. See Lipton, 1991, chapter 9.

53. See Boyd, 1984; Psillos, 1996; and Niiniluoto, 1999.

54. See Leplin, 1982, 1987; and Niiniluoto, 1999.

146   55. On this point see, for example, Sober, 1984.

56. See Salmon, 1984. For an interesting discussion, see Achinstein, 2001. See also Cartwright, 1983, and Mayo, 1986.

57. In his recent paper, Michel Janssen has evaluated a number of important historical instances of what he calls "causal origin inferences," noting that the history of science shows "remarkable continuity" in "basic taxonomies" (see Janssen, 2002). However, for Janssen it is the explanatory power of such inferences that is evidentially important, making then a subspecies of IBE. I disagree that Whewell's view of consilience falls into this camp for, as we have seen, it is the convergence of different lines of (non-explanatory) inference that provides the warrant for the common cause inference. Yet I do agree that this continuity in taxonomy or causal structure is an important aspect of the history of science. Further, unlike Janssen, I believe that such continuities can play a role in arguing for scientific realism.

58. Another benefit of modesty is that, by thinking in terms of causal structure rather than theoretical entities, it allows us to argue in favor of scientific realism while avoiding the morass of theories of reference. Recent attempts to define a scientific realism for theoretical entities while accounting for theory-change have led to somewhat dubious claims, such as the assertion that the term "luminiferous ether" in nineteenth-century optical theories was "really referring" to the "electromagnetic field." See, for example, Psillos, 1999; and Hardin and Rosenberg, 1982. Laudan, 1984, ridiculed such claims.

REFERENCES

Achinstein, P. (2002) "Is There a Valid Experimental Argument for Scientific Realism?" *Journal of Philosophy* 99: 470–95.

———. (2001) *The Book of Evidence*. New York: Oxford University Press.

Boyd, R. (1984) "Realism, Underdetermination, and a Causal Theory of Evidence." *Nous* 7: 1–12.

Buchwald, J. (1989) *The Rise of the Wave Theory of Light*. Chicago: University of Chicago Press.

Butts, R. (1977) "Consilience of Inductions and the Problem of Conceptual Change in Science." In Robert G. Colodny, ed. *Logic, Laws and Life*. Pittsburgh: University of Pittsburgh Press, pp. 71–88.

Cantor, Geoffrey (1983) *Optics After Newton: Theories of Light in Britain and Ireland, 1704–1840*. Manchester: Manchester University Press.

Carrier, M. (1993) "What Is Right with the Miracle Argument." *Studies in History and Philosophy of Science* 24: 391–409.

Cartwright, N. (1983) *How the Lies of Physics Lie*. Oxford: Oxford University Press.

Earman, John (1985) "Concepts of Projectibility and Problems of Induction." *Nous* 19: 521–35.

Ellis, B. (2000) "The New Essentialism and the Scientific Image of Mankind." *Epistemologia* 23: 189–210.

Fine, A. (1996) "Unnatural Attitudes: Realist and Instrumentalist Attachments to Science." *Mind* 95: 149–79.

Fisch, M. (1985) "Whewell's Consilience of Inductions: An Evaluation." *Philosophy of Science* 52: 239–55.

———. (1988) "A Physicist's Philosopher: James Clerk Maxwell on Mathematical Physics." *Journal of Statistical Physics* 51: 309–19.

———. (1991) *William Whewell: Philosopher of Science.* Oxford: Oxford University Press.

Forster, Malcolm R. (1988) "Unification, Explanation, and the Composition of Causes in Newtonian Mechanics." *Studies in History and Philosophy of Science* 19: 55–101.

Friedman, Michael (1983) *Foundations of Space-Time Theories.* Princeton: Princeton University Press.

Hacking, I. (1991) "A Tradition of Natural Kinds." *Philosophical Studies* 61: 109–26.

Hardin, C., and Rosenberg, A. (1982) "In Defense of Convergent Realism." *Philosophy of Science* 49: 604–15.

Harman, P. M. (1998) *The Natural Philosophy of James Clerk Maxwell.* Cambridge: Cambridge University Press.

Harper, W. (1989) "Consilience and Natural Kind Reasoning." In J. R. Brown and J. Mittelstrass, eds., *An Intimate Relation.* Dordrecht: Kluwer, pp. 115–52.

Herschel, J. F. W. (1830) *Preliminary Discourse on the Study of Natural Philosophy.* Facsimile of the original edition, with a foreword by Arthur Fine. Chicago, 1987.

———. (1841) "Whewell on Inductive Sciences." (Review of the *History* and the *Philosophy*). *Quarterly Review* 68: 177–238.

Hesse, M. (1976) "Truth and Growth of Knowledge," in F. Suppe and P. D. Asquith, eds., *PSA: 1976,* vol. 2. East Lansing, MI: Philosophy of Science Association.

Janssen, M. (2002) "COI Stories: Explanation and Evidence in the History of Science." *Perspectives on Science* 10 (4), 457–522.

Kitcher, P. (1993) *The Advancement of Science.* Oxford: Oxford University Press.

Kornblith, H. (1993) *Inductive Inference and Its Natural Ground.* Cambridge, MA: MIT Press.

Laudan, L. (1971) "William Whewell on the Consilience of Inductions." *Monist* 55: 368–91.

———. (1981) "A Confutation of Convergent Realism." *Philosophy of Science* 48: 19–49.

———. (1984) "Discussion: Realism Without the Real." *Philosophy of Science* 51: 156–62.

Leplin, J. (1982) "The Historical Objection to Scientific Realism." *PSA: 1982.* vol. 1. East Lansing, MI: Philosophy of Science Association, pp. 88–97.

———. (1987) "Surrealism." *Mind* 96: 519–24.

Lipton, P. (1991) *Inference to the Best Explanation.* London: Routledge.

Mayo, D. (1986) "Cartwright, Causality and Coincidence." *PSA: 1986.* East Lansing, MI: Philosophy of Science Association, pp. 42–58.

McMullin, E. (1992) *The Inference that Makes Science.* Milwaukee: Marquette University Press.

———. (2001) "The Impact of Newton's *Principia* on the Philosophy of Science." *Philosophy of Science* 68: 279–310.

Morrison, Margaret (1990) "Unification, Realism and Inference." *British Journal for the Philosophy of Science* 41: 305–32.

147

Niiniluoto, I. (1999) "Defending Abduction." *Philosophy of Science* 66 (supplement): S436–51.

Peirce, Charles S. (1865/1982) "Lecture on the Theories of Whewell, Mill, and Comte." In Max H. Fisch, ed., *Writings of Charles S. Peirce: A Chronological Edition*, vol. 1. Bloomington: Indiana University Press, pp. 205–23.

Psillos, S. (1996) "Scientific Realism and the 'Pessimistic Induction'." *Philosophy of Science* 63 (supplement): S306–14.

——. (1999) *Scientific Realism: How Science Tracks Truth.* London: Routledge.

Putnam, H. (1975) *Philosophical Papers*, Vol. I: *Mathematics, Matter and Method.* Cambridge: Cambridge University Press.

——. (1978) *Meaning and the Moral Sciences.* London: Routledge.

Ruse, M. (1975) "Darwin's Debt to Philosophy: An Examination of the Influence of the Philosophical Ideas of John F. W. Herschel and William Whewell on the Development of Charles Darwin's Theory of Evolution." *Studies in History and Philosophy of Science* 6: 159–81.

——. (1976) "The Scientific Methodology of William Whewell." *Centaurus* 20: 227–57.

Salmon, W. (1970) "Bayes's Theorem and the History of Science," in R.H. Steuwer, ed., *Historical and Philosophical Perspectives of Science,* Minnesota Studies in the Philosophy of Science, vol. V. Minneapolis: University of Minnesota Press, pp. 68–86.

——. (1984) *Scientific Explanation and the Causal Structure of the World.* Princeton: Princeton University Press.

Snyder, L. (1997) "Discoverers' Induction." *Philosophy of Science* 64: 580–601.

——. (2002) "Whewell and the Scientists: Science and Philosophy of Science in 19th Century Britain." In *History of Philosophy of Science: New Trends and Perspectives,* M. Heidelberger and F. Stadler, eds. Dordrecht: Kluwer Press, 81–94.

——. (2004) "William Whewell (revised and enlarged version)." *Stanford Online Encyclopedia of Philosophy* (http://plato.stanford.edu/Whewell).

——. (forthcoming) *Reforming Philosophy: A Victorian Debate on Science and Society.* Chicago: University of Chicago Press.

Sober, E. (1984) "Common Cause Explanation." *Philosophy of Science* 51: 212–51.

——. (1989) "Independent Evidence about a Common Cause." *Philosophy of Science* 56: 275–87.

Van Fraassen, B. (1980) *The Scientific Image.* Oxford: Oxford University Press.

——. (1985) "Empiricism in Philosophy of Science." in P. M. Churchland and C. A. Hooker, eds., *Images of Science.* Chicago: University of Chicago Press.

Whewell, W. (1828) *An Essay on Mineralogical Classification.* Cambridge: J. Smith.

——. (1833) *Astronomy and General Physics with Reference to Natural Theology.* London: John W. Parker.

——. (1847) *The Philosophy of the Inductive Sciences, Founded Upon Their History,* 2nd edition (two volumes). London: John W. Parker.

——. (1857/1873) *The History of the Inductive Sciences, from the Earliest to the Present Time,* 3rd edition, with additions; reprinted in 1873, in 2 volumes. New York: D. Appleton and Co.

——. (1858) *Novum Organon Renovatum.* London: John W. Parker.

——. (1860) *Philosophy of Discovery.* London: John W. Parker.

Worrall, J. (1989) "Structural Realism: The Best of Both Worlds?" *Dialectica* 43: 99–124.

PART II

SCIENTIFIC APPLICATIONS

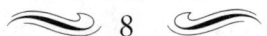

8

EVIDENCE FOR TRANSMUTATION IN SEVENTEENTH-CENTURY ALCHEMY

Lawrence M. Principe

It may seem strange to have chosen alchemical transmutation as a chapter topic in a book on scientific evidence. In popular conceptions, alchemy is lamentably labeled with the problematic title of pseudo-science, and thus seems to be far removed from any relevance to the topic of scientific evidence, save perhaps as a counter-example. This state of affairs is due largely to the fact that alchemy has been for a long time used quite uncritically as a foil to modern science. Yet historical studies carried out over the past thirty years have been displaying the error and artificiality of such a perspective, and alchemy is regaining—at least among historians of early modern science—not only a more unbiased hearing, but also its due place in the genealogy of modern science.[1]

In the course of this chapter I hope to continue this reevaluation, as well as simultaneously contributing further illustrations to our discussions of evidence by outlining the ways in which various kinds of evidence were marshaled, evaluated, and deployed in favor of alchemical transmutation. In the latter half of the chapter I also note the curious fact—and one I think important to account for in theories of evidence—that the eventual dismissal of alchemical transmutation as a real phenomenon had little or nothing to do with scientific evidence, nor can it be traced to a shift in or development of the epistemic background.

It is wise to begin with two definitions-first of this chapter's approach, and then of its subject. The approach of this chapter is primarily a historical one, even though it does posit some questions and problems of a philosophical nature as regards the meaning, types, and definition of scientific evidence. As for the subject, by "alchemical transmutation" I mean the conversion of base metals (mercury, lead, tin, copper, and iron) into gold or silver by a process called projection. Projection is a means of metallic transmutation requiring the use of the Philosophers' Stone, also called the Elixir. The Philosophers' Stone was a chemical substance which was to be prepared by a lengthy laboratory process requiring up to a year for its completion. A small particle of the Stone, often wrapped in wax, was to be thrown onto a quantity of molten lead or boiling mercury. The mixture was heated strongly for a few minutes, in which time the entire mass of metal would be transmuted into pure gold.[2] This process of projection takes its name from the Latin word *proiecere,* literally to throw or cast upon, in reference to the casting of a particle of the Philosophers' Stone onto hot metals. The amount of metal transmuted depended upon the quality of the Stone used, but generally ranged from ten to ten thousand times the weight of the Stone.

Aspiring transmuters of metals (or chrysopoeians, from the Greek words *chryson poiein,* "to make gold") believed that a very few of their colleagues past and present (called adepts) had succeeded in preparing the Stone in their laboratories. The difficulty was, of course, that the actual procedure for making the Stone was to be kept highly secret, as was the fact that one might possess it. This is what one would expect of knowledge which was at once both so valuable and so potentially dangerous to the owner and to the economic and political stability of states dependent upon the value—hence the scarcity—of precious metals. Nevertheless, throughout the early modern period there was a great deal of discussion in both public and private writings regarding the proper constituent materials and their laboratory treatment. Moreover, the physical properties of the Stone were also well-known: It was supposed to be red in color, fusible as wax, extremely dense, and able to penetrate the metals as oil does paper.

Now after this brief introduction to the alchemical summum bonum, the reader may well be wondering about the bases upon which so many generations of Stone-seeking alchemists rested their firm conviction that the Stone was, in fact, a real and preparable substance with potentially stupendous transmuting powers. Although some sneering positivists have claimed in the past that there was little if any evidence for this belief, only avarice and daydreaming imagination, there were in fact several varieties of evidence that testified in favor of projective transmutation and were wielded in that way by chrysopoetic apologists of the early modern period.

Virtually from the time in which the subject first entered the Latin West from the Islamic world in the twelfth century, alchemy was a controversial topic. Early topics of debate involved the possible ways of undertaking the preparation of the Stone and how it operated, as well as questions about its licitness (for example, theologians queried whether it was moral to sell alchemically prepared gold in the same way as natural gold), and of course about the real existence or preparability of al-iksir, the Philosophers' Stone. These debates persisted down to the seventeenth century, and it is from this latter period—the full flowering of European alchemy—that the following examples are taken.

The hypothesis for which we are going to evaluate the evidence, then, is That the Philosophers' Stone exists as a preparable substance which can transmute metals by projection.

The first point to stress is that transmutation was supported by the most sophisticated chymical theories of the day. Accordingly, belief in the possibility of transmutation was rendered reasonable because it was a consequence of prevailing chymical theories. Thus, belief in transmutation did not exist as a notion unrelated to, separate from, or contrary to a coherent, serious, and widely supported system of natural philosophical thought. This relationship distinguishes belief in transmutation from belief in, say, the existence of the Loch Ness Monster. The latter is not predicted, entailed, or rendered more likely true than false by any set of coherent scientific theories.

Using the categorization of types of evidence advocated by Achinstein, such evidence or theoretical backing for the Stone is objective evidence relativized to the prevailing epistemic situation (ES-evidence). That is, assuming the truth of prevailing chymical theories, the possibility of metallic transmutation is predicted by them.[3] For example, the most widespread matter theory envisioned metals as composite bodies composed of the same basic ingredients, which—depending on one's particular theoretical commitments—were either two, mercury and sulphur, or three, with the addition of salt. Metals were formed naturally in the bowels of the earth when the vapors of mercury and sulphur combined underground. Gold resulted when very pure mercury and sulphur mixed in precise proportions, but when the two combined in the wrong proportion and together with impure earthy matter, one of the other metals was formed. Thus, since all metals are composed of the same ingredients but in differing proportions and grades of purity, transmutation is possible by purification and adjustment of proportions.[4] The Philosophers' Stone acts by burning away superfluities and impurities.

More strictly Aristotelian thinkers preferred the language of matter, form, and qualities, and this too was consonant with transmutation. One such writer, Gaston Duclo, argued according to Scholastic terms that gold was different from lead in regard to its qualities, the primary qualities being hot-cold and dry-wet. Therefore, new qualities need to be imposed on a base metal to produce gold, and the Philosophers' Stone does exactly that. It contains the hot and dry qualities of gold to an intensified degree, so that transmutation is nothing other than the true mixture of the intense qualities of the Stone with the remiss qualities of the base metal. Thus, so long as the proportion is correct, gold will naturally result from the combination.[5] There were yet other theories of material composition, including particulate or atomistic theories, and the possibility of metallic transmutation was rendered reasonable by them as well.[6]

Now let us turn to some seventeenth-century chymical apologetics to illustrate what other sorts of evidence were brought to bear on the question of the reality of the Stone and of projective transmutation. One valuable and surprising source which has only recently been brought to light is the unpublished *Dialogue on the Transmutation and Melioration of Metals,* written around 1680 by none other than Robert Boyle himself. The *Dialogue* is set as a meeting of the Royal Society, and takes the form of a debate between those who reject the reality of the Philosophers' Stone—called the anti-Lapidists—and those who support its reality—the Lapidists (from the Latin *lapis,* for Stone). Boyle himself weighs in strongly on the side of the Lapidists. After responding to specific criticisms leveled by his anti-Lapidist opponent, Boyle's spokesman states that transmutation by the Philosophers' Stone can be first "countenanced by things Analogicall in nature or Art" and then "prov'd, by particular Histories & other testimonies."[7] These two sorts of evidence are the usual participants in alchemical debates of the seventeenth century, and they bear a closer look.

Argument by analogy was an important source of evidence—though of a weaker sort—for alchemical writers, and Boyle reprises some of the classical examples in his *Dialogue.*[8] In regard to the seemingly incredible transformations of properties which the Stone can effect quickly on a large quantity of metal, there are analogous examples drawn from common experience. For example, he notes how a small proportion of rennet can curdle a much greater quantity of milk, or how a small fire can multiply itself through a vast amount of kindling, or how one part of leaven mixed with ten parts of flour converts everything into leaven, or finally, how a few drops of vinegar added to a jug of wine soon convert all of it into vinegar.

But Boyle does not claim that these "things Analogicall in nature and art" provide evidence strictly speaking, but rather only that they countenance the existence of what he trying to prove. By providing analogous precedents to the action of the Stone, they provide him with evidence in favor of the non-impossibility rather than of the reality of the Stone and its powers. In order to prove more confidently the real existence of the Stone, Boyle—like most other alchemical apologists of the seventeenth century—turns instead to empirical data as the best sort of evidence. In this case, he cites the empirical data of experiential accounts, in other words, eyewitness accounts of transmutation by projection.

Whereas theoretical considerations played a major role in alchemical apologetics throughout the history of alchemy in the Latin West, by the seventeenth century accounts of transmutations had risen to an equal or even greater importance. Indeed, we can point to an entire genre of alchemical writings which may be termed "transmutation histories." These transmutation histories were published either singly or as collections, and the apologists' arsenal was increasingly stockpiled with them throughout the century. Many involve stories of traveling adepti who performed projection privately before one or more aspiring alchemists or alchemical skeptics. Although some such accounts are so romanticized or so outlandish that they rarely fail to provoke a wry smirk from the modern reader, at least an equal number are painstakingly precise, noting exact times, places, persons (often of recognized name and credibility) in attendance, the quantity of gold or silver produced, and so forth.[9] Some examples take place at a court or assembly, and were sometimes commemorated by the striking of coins or medallions, often from the transmuted metal itself. By the end of the century, a sufficient number of such coins had been minted that an entire dissertation was written about them.[10]

155

Several accounts of the witness of projection achieved great notoriety. For example, there is the account published in 1667 by Johann Friedrich Helvetius (1625–1709), physician to the prince of Orange.[11] According to the report, on the afternoon of 27 December 1666, a stranger visited Helvetius at his house in The Hague. Helvetius had written skeptically about the validity of alchemy, and the visitor engaged him in conversation on this topic. After some discussion, the visitor took out a small ivory box and showed Helvetius its contents—three heavy lumps of a glassy substance—which he claimed to be the Philosophers' Stone in sufficient quantity to produce 20 tons of gold. In due course, the stranger gave Helvetius a fragment of the Stone "smaller than a rapeseed." After the man had left, Helvetius melted down some lead, cast the particle of the Stone upon it according to the directions he had been given, and successfully transmuted the lead into gold. Helvetius then took the transmuted metal to be assayed by the mint-master, who not only found it to be pure gold, but also, when he melted the sample with silver in order to assay it, found that the alchemical gold actually transmuted some of the added silver, producing a 33% increase in the amount of gold. That result was produced, as Helvetius suggests, by an "excess of tincture" in the alchemical gold. (In modern terms, the lead was the limiting reagent in the reaction, so some unreacted Philosophers' Stone was left in the transmuted metal. Upon being mixed with silver at the assay-master's, this residual Stone acted upon the added silver, producing more gold.) This account became widely known across Europe, and intellectuals sought to verify the facts. Benedict Spinoza, for example, traveled to visit both Helvetius and the assayer.[12]

Boyle's *Dialogue* presents several accounts of transmutation which had been privately relayed to Boyle. But most strikingly, it also contains a detailed account of Boyle's own eyewitness account of the Stone and projective transmutation. According to this document, Boyle, while visiting a "forraigne doctor of Phisick," is introduced to a certain foreigner who attempts to show him a method whereby lead can be turned into a metallic substance that is liquid at room temperature, like mercury. Boyle's servant is sent to obtain lead and crucibles for the experiment, but the experiment miscarries when the crucible is upset, spilling its contents into the fire. The foreigner then agrees to demonstrate another experiment, which Boyle mistakenly assumes to be a mere repetition of the miscarried one.

> Then the Lead being strongly melted the Traveller opened a small peice of folded paper wherein there appear'd to be some grains but not very many of a powder that seemd somewhat transparent almost like exceeding small Rubies & was of a very fine & beautifull red. Of this he tooke carelessly enough, & without weighing it upon the point of a knife as much as I guest to be about a graine or at most betwixt one graine & two & then presenting me the haft of the knife he told me that I might if I pleas'd cast in the powder with my owne hand. But the fire by that time burning feircly & I haveing that morning had an accidentall indisposition in my eyes found them so Dazelld & offended by the glowing fire when I came very near it, that imagineing that the Experiment to be made was but the same that lately succeeded so ill I was unwilling to hurt my eyes & apprehensive lest the experiment might miscarry in my hands & therefore restoreing the knife to the Traveller I desired him to cast in the powder himselfe which he did whilst I stood by & lookd on.[13]

After covering the crucible and heating it strongly for fifteen minutes, they take the crucible out of the fire and let it cool. Then,

> the Crusiple haveing been kept till it was cool enough to be managed without doeing harme we remov'd it to the window where in stead of runing Mercury I was surprizd to find a solid Body & my surprize was

increasd when the Crusiple being inverted tho yett a little hott the Mass that came out & still retaind the figure of the lower part of the vessell apear'd very yellow & when I took it into my hand felt to my thinking manifestly heavier then so much Lead would have done. Upon this, turning my eyes with a somewhat amazed look upon the Travellers face he smild & told me he thought I had suffitiently understood what kind of experiment that newly made was design'd to be.[14]

Boyle then took the yellow metal, and after subjecting it to all assays and tests, found it indeed to be pure gold. A short time afterward, a friend of Boyle's, whom we can identify with reasonable certainty as Edmund Dickenson, professor of medicine at Oxford and first physician to the King, comes to tell him that he met the same traveler in Oxford. The traveler showed him not one but two transmutations, one upon lead, and the other upon copper, "and in this last," writes Boyle, "the Physitian for fuller satisfaction would needs have the operation try'd on some of our English Copper farthings that he took out of his owne Pockett, which tho much more Difficulty melted then the Lead had been, were no less really transmuted into Gold."[15]

This evidence proved sufficient for Boyle, and he later told his confessor, the bishop Gilbert Burnet, that this incident gave him "convincing satisfaction in that matter."[16] Indeed, both Boyle and Burnet testified to the reality of transmutation before Parliament in order to have the English law forbidding transmutation repealed in 1689. The Royal Society Journal Book records that "Inducement to the Parliament to Repeal the Act of Henery 4th against Multiplication of Metalls was from the Testimony of Mr Boyle and the Bishop of Salisbury who affirm to have seen projection, or the transmutation of other metals into Gold."[17] Just as this eyewitness of projection provided evidence which convinced Boyle, so the account of it is used as evidence to convince others who would read the *Dialogue*.

At this point I would like to remark upon the difference in the sort of evidence that comes from experiences that cannot be repeated at will. On the one hand, we have phenomena that are under our control—we can always drop heavy objects and plot their descent, or prepare various sorts of tubes to study cathode rays—whether to make our own fresh observations or to verify someone else's. But what of phenomena that are beyond our power to bring about? There exist more than one of this sort. Supernovae are quite rare events that are beyond our ability to cause to happen, but when they do happen they are fairly accessible, owing to the many observatories and observers. Other rare natural phenomena—such as the predicted decay of a proton—should occur at statistically regular intervals, and even if the apparatus needed to observe them is unique and the observers few, if the observation is carried on long enough, one should witness an event.

Instances of transmutation, however, are not demandable or widely observable, nor do they occur on a regular timetable. A charitably disposed adept has to choose you as observer. You could not find the secretive adept yourself, stand in one place long enough for him to pass by, compel him to show you the Stone, or enforce that he do so on demand before any number of future inquirers. Some natural phenomena are like this too—for example, a lone biologist being in the right place at the right time to witness by chance a peculiar behavior of a rare species, or the erratic, unpredictable, and debatable appearance of such things as ball lightning. Seen by a few, at irregular and unpredictable intervals, their value as evidence is subjected to a substantial range of influences—social as well as intellectual—which would seem to need to be accounted for in our accounts of scientific evidence.

158

In terms of prediction and explication by chymical theory, analogy, and eyewitness testimony there was sufficient evidence of the reality of projective transmutation to render belief in it quite reasonable. Many serious and learned intellectuals of the seventeenth century did in fact accept its existence, and argued in favor of it. But within a period of a single generation, from the 1690s to the 1730s, belief in projective transmutation disappeared almost completely from learned culture. Although it is no surprise to have long-held scientific beliefs overturned within a relatively short period of time, the case of alchemy remains peculiar (although perhaps not unique) because it has proven difficult for historians to identify the true cause for its dismissal. Of course, those who lean toward a triumphalist view of the history of science have paid little attention to the details of the demise of alchemy, since in their view it was only "a matter of time" before the inexorable progress of science rejected it as a false belief more or less unproblematically.

But such a view implies that some compelling piece of evidence actually proved the falsity of our hypothesis. The advance of chymical and atomic theory would seem to be a reasonable place for such evidence to emerge, but if we examine chymical theory of the late seventeenth and early eighteenth centuries, we find no new theory or experiment whose persuasive force could render belief in projective transmutation unreasonable. On the contrary, the foremost chymical theorist and chief chymist of the Parisian Académie Royale des Sciences, Wilhelm Homberg (1652–1715), was convinced of the truth of metallic transmutation and actively sought the Stone in his laboratory.[18] So we cannot point to a "crucial experiment" or an epistemic change in the scientific background information that would undercut the validity of the existing evidence or the reasonableness of belief in the hypothesis.

If we examine the development of alchemical debates, however, we can get at one possible explanation, for there was a significant change in the terms of the arguments at just about the time alchemy began to falter. Recall that debates over the validity of transmutation had been part and parcel of alchemy for 500 years. In the late sixteenth century, Thomas Erastus published a violent denunciation of transmutation. His volley was returned in a short treatise by Gaston Duclo, who met and refuted Erastus methodically point by point.[19] At about the same time, Nicolas Guibert published an equally vigorous denial of alchemy, which was met by a ponderous tome of invective lauched from the Teutonic pen of Andreas Libavius.[20] At the middle of the seventeenth century, the Jena professor Werner Rolfinck derided the Stone as a "chymicall non-entity," and the Jesuit Athanasius Kircher claimed that the Stone was false and diabolical. Both received lengthy refutations from equally learned pens—from the professor and polymath Daniel Georg Morhof, and from Gabriel Clauder, Salomon von Blauenstein, Johann Joachim Becher, and several others.[21]

But something changed in the final years of the seventeenth century. In 1679, the French chymical textbook writer and pharmacist Nicolas Lemery condemned all accounts of transmutation as mere fraud. No theory new or old rendered them unlikely, no logic or new evidence was marshaled against his opponents; he merely stamped them all with the label of cheats.[22] Etienne-François Geoffroy did the same in his 1722 paper "Some cheats concerning the Philosophers' Stone," published in the august pages of the annual memoirs of the Parisian Academy of Sciences.[23] Both of these publications went largely unanswered. Apparently, those who continued to support the reality of the Stone fell silent. In the 1680s Boyle demurred from publishing his *Dialogue* and most of his other alchemical writings; Isaac Newton, whose alchemical pursuits have become a cause célèbre, kept his interests and activities hidden; and Wilhelm Homberg, although he published some of his transmutational experiments in the first decade of the eighteenth century, did so only in veiled manner without linking them explicitly to transmutation, and he never ventured into print to refute the claims of Lemery, his fellow Academicien.

Thus, toward the end of the seventeenth century alchemical apologists retired from the field, not under a hail of argument or contrary evidence, but probably, as it seems to me at this point, in fear of the social and moral opprobrium with which projection had been loaded by the likes of Lemery. The costs for the public defense of projective transmutation had simply been made too high. Alchemy thus seems to have died of scorn and innuendo, rather than on the scientific battlefield under an assault of evidence and reason.

This account of transmutation's demise remains incomplete, however, for the cheating practices described by Lemery and Geoffroy were nothing new. In fact, many of the specific examples they use are recycled from the *Examen fucorum pseudochymicorum detectorum* of 1617, written by the supporter of transmutation, Michael Maier. Maier had listed these methods of fraud so that the lucky observers of projections would be able to detect any tricks and could thus identify the true transmutations with a greater level of certainty. Indeed, the existence of alchemical frauds was well enough known in the fourteenth century to be used to comic effect in Chaucer's "Canon's Yeoman's Tale" and several later pieces of literature and theatre.

So why did the same old, well-known accusations and descriptions of fraudulent practices suddenly prove fatal to transmutatory alchemy at the end of the seventeenth century? Undoubtedly there exists a whole constellation of reasons. One was certainly the professionalization of chemistry beginning around this time, perhaps instanced most obviously in the enshrinement of chemistry as one of the scientific disciplines supported (now for the first time alongside such old favorites as astronomy and physics) at the Académie Royale des Sciences, by the French Crown. At this juncture, when chemists and chemistry were becoming publicly visible and socially respectable for the first time, the problem of the association with transmuters and the potential accusations of fraud was particularly delicate, and it led to a separation of gold-making from the rest of chymistry. This segregation is witnessed by the separation of the words alchemy and chemistry, which had previously been largely synonymous, but took on nearly their modern distinct definitions by 1720 at the latest.[24] Under such social and institutional conditions, it was undoubtedly strategically advisable for chemists nervous about their status (as Lemery, for example, most certainly was) to distinguish themselves unambiguously from a subject that was not only risky, but also tainted with fraudulent hangers-on.[25] Likewise, supporters of transmutation (often with public persona and status like Boyle and Homberg) found it wiser under such external circumstances to continue their researches quietly, without engaging in public debate.

In addition, the increased acceptance of chemical medicaments and methods in medicine that also occurred around 1700 meant that some who might otherwise have devoted themselves to the quest for rare chymical arcana such as the Philosophers' Stone now had a more easily accessible, more socially respectable, and more likely lucrative realm in which to exercise their chymical interests and talents, rather than hazarding themselves in the potentially disastrous quest for the elusive Stone. Thus, not only did alchemy's champions retire from the field, but their ranks were not regularly replenished as they would have been a generation earlier.

Whatever the exact combinations of influences, what I mean to emphasize is that what had been the strongest evidence for our hypothesis—eyewitness testimony—was rendered automatically suspect by external conditions. The testimonies and the testifiers were ridiculed as dupes or frauds, and consequently such empirical evidence in effect weakened rather than strengthened the proposition by being viewed as evidence of fraud, rather than as evidence of a real phenomenon. Significantly, this inversion occurred without the support of any new experiment, or change of theory or epistemic background. Indeed no sound, theoretical, scientific evidence against the possibility of metallic transmutation by chemical means was supplied until the early twentieth century, when the discovery of atomic structure proved that what we had been calling elements only provisionally and by convention since the time of Lavoisier and Dalton were really distinct and non-interconvertible elements (at least by chemical means) at the atomic level. But for the demise of transmutational alchemy, we must point instead to a set of historically contingent social, cultural, and professional factors that came together around 1700. They dealt the death blow within a generation and without discernible debate to a belief whose reasonableness had withstood assault for the previous 500 years.

My point is, then, that the evaluation of evidence for the alchemical hypothesis was dependent upon far more complex factors than could be reduced to a straightforward logical analysis of intellectual factors divorced from the historical context. Of course Lemery, for example, had his reasons for denying transmutational alchemy, as outlined earlier, but these were not scientific evidence against the proposition. The details of the case of projective transmutation and its sudden rejection should be accounted for in theories of evidence that endeavor to describe actual scientific practices. It is not sufficient to fall back upon anti-alchemical prejudice or discredited positivist notions that alchemical transmutation was "doomed to be rejected" as a falsehood, for it is undoubtedly the case that there are currently accepted scientific notions, replete with good evidence, that shall in due course themselves be rejected. There is no solid ground for differentiating these notions from the transmutational hypothesis—both were reasonable and widely supported by leading intellectuals and the best scientific theories of their day. To consider a piece or body of evidence as unsatisfactory based not upon its historical context, but rather in hindsight based upon subsequent scientific development, would imply that the accurate and final assessment of evidence could be accomplished only by a being possessing the sum total of all knowledge. This would reduce the number of qualified philosophers to at most one, and ironically, to the One who needs a theory of evidence the least.

In conclusion, it is safe to say that an intellectual of 1600 or 1650 had good theoretical and experimental evidence in favor of metallic transmutation. Alchemy had as much a claim to scientific sophistication as any such endeavor. Moreover, the demise of alchemy did not come about as an unproblematic and rational consequence of scientific development; we cannot point to a theoretical or experimental negation of its transmutatory claims. Instead, factors that are historically contingent—based upon professionalization of disciplines, social valuations, and so forth—were the key factors, with firm scientific evidence appearing only about two centuries after metallic transmutation by means of the Philosophers' Stone was largely laid to rest and the question of its possibility had become moot.

NOTES

1. The work on uncovering the alchemical preoccupations of such chief figures of the Scientific Revolution as Sir Isaac Newton and Robert Boyle is one example. For an entrée into recent reevaluations of alchemy, see the pair of papers: Lawrence M. Principe and William R. Newman, "Some Problems in the Historiography of Alchemy," pp. 385–434 in *Secrets of Nature: Astrology and Alchemy in Early Modern Europe*, ed. William R. Newman and Anthony Grafton (Cambridge, MA: MIT Press, 2001), and William R. Newman and Lawrence M. Principe, "Alchemy vs. Chemistry: The Etymological Origins of a Historiographic Mistake," *Early Science and Medicine* 3 (1998): 32–65.

2. Some readers may be surprised that the Philosophers' Stone is a physical, chemical substance, since many popular treatments of alchemy emphasize the "spiritual" dimension of alchemy, thus distinguishing it strongly from more familiar chemistry. This "spiritual interpretation" is, however, predominantly a product of the Victorian occult revival (see Principe and Newman, "Historiographical Problems").

3. Peter Achinstein, *The Book of Evidence* (N.Y.: Oxford University Press, 2001).

4. It must be noted that this first type of evidence for the Stone is not particularly *strong*, because while the chymical theories do predict the theoretical possibility of transmutation, they do not guarantee that the process can in fact be effected practically in the laboratory. This distinction was exploited by some critics of transmutation. See for example, the summary of critiques given by Geber in the *Summa perfectionis*, ed. and trans. William R. Newman (Leiden: Brill, 1990).

5. Lawrence M. Principe, "Diversity in Alchemy: The Case of Gaston "Claveus" DuClo, a Scholastic Mercurialist Chrysopoeian," pp. 181–200 in *Reading the Book of Nature: The Other Side of the Scientific Revolution*, ed. Allen G. Debus and Michael Walton (Kirksville, MO: Sixteenth Century Press, 1998).

6. On particulate matter theory and transmutation, see William R. Newman, "The Corpuscular Transmutational Theory of Eirenaeus Philalethes," pp. 161–82 in *Alchemy and Chemistry in the Sixteenth and Seventeenth Centuries*, ed. Piyo Rattansi and Antonio Clericuzio (Dordrecht: Kluwer, 1994).

7. Boyle's "lost" *Dialogue* is analyzed in Lawrence M. Principe, *The Aspiring Adept: Robert Boyle and His Alchemical Quest* (Princeton: Princeton University Press, 1998); the full text and translation are published on pp. 233–95.

8. This sort of evidence is *subjective evidence* according to Achinstein's taxonomy in *The Book of Evidence*.

9. An early example is Ewald van Hoghelande, *Historiae aliquot transmutationis metallicae . . . pro defensione alchymiae contra hostium rabiem* (Cologne, 1604). A much larger, and very late, collection is Siegmund Heinrich Güldenfalk, *Sammlung von mehr als hundert wahrhaftigen Transmutationgeschichten* (Frankfurt and Leipzig, 1784).

10. Samuel Reyher, *Dissertatio de nummis quibusdam ex chymico metallo factis* (Kiel, 1690).

11. Johann Friedrich Helvetius, *Vitulus aureus,* pp. 815–63 in *Musaeum hermeticum* (Frankfurt, 1678; reprint ed. Graz: Akademische Druck, 1970).

12. *Spinoza Opera im Auftrag der Heidelberger Akademie der Wissenschaften,* ed. Carl Gebhardt, 5 vols, (Heidelberg, 1925); vol. 4: *Epistolae,* pp. 196–97.

13. Robert Boyle, *Dialogue on Transmutation,* in Principe, *Aspiring Adept,* p. 265; the original document is Royal Society Boyle Papers, vol. 7, fols. 160–63v.

14. Ibid., p. 266.

15. Ibid., p. 268.

163

16. "Burnet Memorandum" printed in Michael Hunter, *Robert Boyle By Himself and His Friends* (London: Pickering, 1994), p. 30.

17. See Michael Hunter, "Alchemy, Magic, and Moralism in the Thought of Robert Boyle," *British Journal for the History of Science* 23 (1990): 387–410, on p. 405.

18. Lawrence M. Principe, "Wilhelm Homberg: Chymical Corpuscularianism and Chrysopoeia in the Early Eighteenth Century," pp. 535–56 in *Late Medieval and Early Modern Corpuscular Matter Theories,* ed. Christoph Lüthy, John Murdoch, and William Newman (Leiden: Brill, 2001).

19. Thomas Erastus, *Explicatio quaestionis famosae illius, utrum ex metallis ignobilibus aurum verum et naturale arte conflari possit,* in *Disputationum de nova Philippi Paracelsi medicina pars altera* (Basel, 1572). On Erastus, see Charles D. Gunnoe, Jr., "Thomas Erastus and His Circle of Anti-Paracelsians," pp. 127–148 in *Analecta Paracelsica,* ed. Joachim Telle (Stuttgart: Franz Steiner Verlag, 1994); Gaston Duclo, *Apologia argyropoeiae et chrysopoeiae,* (Nevers, 1590).

20. Nicolas Guibert, *Alchymia ratione et experientia . . . impugnata et expugnata* (Strasbourg, 1603); Andreas Libavius, *Defensio et declaratio perspicua alchymiae transmutatoriae, opposita Nicolai Guiperti* (Ursel, 1604).

21. Werner Rolfinck, "Non-entia chymica," in *Chemia in artis formam redacta* (Jena, 1661); Athanasius Kircher, "Dissertatio de lapide philosophorum" (extracted from his *Mundus subterraneus* of 1665), in J. J. Manget, ed., *Bibliotheca chemica curiosa,* 2 vols. (Sala Bolognese: Arnaldo Forni, 1976; reprint of 1702 Geneva edition), I: 38–112; Daniel Georg Morhof, *Epistola ad Joelum Langelottum de transmutatione metallorum* (Hamburg, 1673), on this treatise see Lawrence M. Principe, "D. G. Morhof's Analysis and Defence of Transmutational Alchemy," pp. 139–53 in *Mapping the World of Learning: The Polyhistor of Daniel Georg Morhof,* ed. Françoise Waquet (Wiesbaden: Harrassowitz, 2000); Gabriel Clauder, *Dissertatio de tinctura universalis* (Altenburg, 1678); Salomon von Blauenstein, *Interpellatio brevis ad Philosophos pro lapide philosophorum,* I: 113–19 in *Bibliotheca chemica curiosa;* and Johann Joachim Becher, *Experimentum chymicum novum . . . loco . . . responsi ad D. Rolfincii,* pp. 281–346 in *Physica subterranea* (Leipzig, 1738; originally published 1671).

22. Nicolas Lemery, *Cours de chemie,* 3rd ed. (Paris, 1679), pp. 57–61. On Lemery see Michel Bougard, *La chimie de Nicolas Lemery* (Turnhout: Brepols, 1999); and on his denunciation of alchemy, Newman and Principe, "Alchemy vs. Chemistry," pp. 59–61, and John C. Powers, "'Ars sine Arte': Nicholas Lemery and the End of Alchemy in Eighteenth-Century France," *Ambix* 45 (1998): 163–89.

23. Etienne-François Geoffroy, "Des supercheries concernant la pierre philosophale," *Mémoires de l'Académie Royale des Sciences* 24 (1722): 61–70.

24. See Newman and Principe, "Etymological Origins."

25. As a Protestant, Lemery's social position was highly unstable in the years leading up to the revocation of the Edict of Nantes; indeed, he eventually lost his laboratory and his right to practice in Paris.

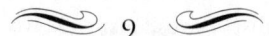

AGENCY AND OBJECTIVITY IN THE
SEARCH FOR THE TOP QUARK

Kent W. Staley

On April 26, 1994, the Collider Detector at Fermilab Collaboration announced the submission of a paper to Physical Review D bearing the title "Evidence for Top Quark Production in p̄p Collisions at √s = 1.8 TeV." The word "evidence" reappears in the conclusions at the end of the paper: "The data presented here give evidence for, but do not firmly establish the existence of, tt̄ production in p̄p collisions at √s = 1.8 TeV" (Abe et al. 1994: 3023). What sort of a claim is this?

Some philosophers interpret such evidence statements, in some contexts, as statements of objective facts that obtain independently of what anyone does or should believe. Others (personalist Bayesians, for example) take such statements to express something primarily about what some person or persons either do or should believe the facts are, given their other beliefs. One advocate of the former viewpoint is Peter Achinstein.

In *The Book of Evidence,* Achinstein defends the centrality to scientific reasoning and practice of what he calls *potential evidence.* According to his account, if some fact E is potential evidence for a hypothesis H, then E is a good reason to believe H, even if no one knows or even believes that E is evidence for H, even if no one actually believes H for the reason that E, and even if no one knows or even believes that E is the case. This concept of evidence does not require that H is true. According to Achinstein, "E is *veridical evidence* for H" is true just in case E is potential evidence for H, and H is true.[1] Crucial to Achinstein's way of conceiving the objectivity of evidence is that potential evidence is not relativized to any person or group of persons, or to any belief set or epistemic situation.[2] Achinstein is thus committed to the following principle:

> β: *If E is potential evidence that H, then E is a good reason to believe H, independently of any person or epistemic situation.*

Another advocate of objectivity in the theory of evidence is Deborah Mayo. In her *Error and the Growth of Experimental Knowledge,* she contrasts the objectivity of her concept of error statistical evidence with the pernicious subjectivity of personalist Bayesianism. Mayo is less systematic than Achinstein in specifying what such objectivity amounts to, but at least two main points are central to her understanding of the objectivity of error statistical evidence. First, evidential judgments are corrigible, and hence are subject to error. She endorses Henry Kyburg's insistence that the possibility of error is a "touchstone of objectivity" (Mayo, 1996: 83). Second, evidential relationships supervene on the error probabilities of testing procedures, which are not a matter of opinion, but are facts about long run relative frequencies.

My question is whether the concept of objectivity codified in Achinstein's principle is compatible with Mayo's error statistical theory of evidence. The motivation for undertaking this investigation is that the error statistical theory strikes me as a promising account of scientific evidence with strong connections to scientific practice. Yet the sense in which it is objective requires clarification. Achinstein's careful articulation of different concepts of evidence with distinct types of objectivity provides a useful starting point for such an investigation, although Achinstein employs these concepts in defense of his own theory of evidence, which I will not here discuss.

I wish to explore two sources of difficulty facing an Achinsteinian understanding of error statistical objectivity. The first concerns the role of the experimental agent in performing the tests that are central to the error statistical theory of evidence. This difficulty will prove to be merely apparent. However, the resolution of the challenge from experimental agency will highlight a deeper problem with β concerning the nature of reasons, and whether they can be the kinds of things Achinstein claims. I will offer in place of β an alternative formulation of the sense in which error statistical evidence provides objective reasons to believe empirical claims.

THE ERROR STATISTICAL THEORY OF EVIDENCE

In Mayo's account, experimental evidence for an empirical claim arises from a process of testing, and the evidential status of the results of experiment depends upon certain characteristics of the testing procedure.

Specifically, suppose that we subject our hypothesis H to a testing procedure T that yields a result E. Then, on Mayo's account, E is evidence for
H just in case:

(1) E fits H, and

(2) H passing T with E constitutes a severe test of H.

Satisfying (1) is a matter of degree. The point of requirement (1) is that H
and its competitors are not simply a means of deducing some disorganized
collection of predictions; rather, some possible results should be "closer," in
some—typically probabilistic—sense, to what H leads us to expect than
others; E should be a result that is close to what H leads us to expect.
Requirement (2) is to be understood as equivalent to the following probability claim:

(2′) Prob (H passes T with a result such as E | ¬H) is very low,

where "a result such as E" should be understood as "a result that fits H as
well as or better than E." If H is being tested against a compound hypothesis—
one that comprises more than one distinct alternative to H—then the
probability in (2′) will not be well-defined, and we must instead evaluate
the probability of H passing T with E under specific alternatives to be
ruled out. In any case, since "very low" admits of degrees, so does the satisfaction of requirement (2).

The probabilities referred to in this account are intended by Mayo to be
objective frequentist probabilities. E being evidence for H depends on the
probability distribution for the outcome space (including E) determined
by H, and on the probabilities of H's passing T with an outcome such as E
determined by the alternatives against which H is tested. Hence, under the
error statistical theory, one can make the following claim: *evidential status
supervenes on strictly objective facts.*

Though this claim seems straightforward at first, it is worth pausing
to probe more deeply the conceptual elements in this account. Some
issues will then arise that challenge at least some common conceptions of
objectivity.

ARE TESTS OBJECTIVE? A PROBLEM FROM THE SEARCH
FOR THE TOP QUARK

What counts as evidence on the error statistical analysis is not a fact of just
any sort, but a fact regarding the outcome of a testing procedure. Whether

E is evidence for H is not a matter of E alone, but of the testing procedure that produced E as an outcome.

Or, to put it another way: In an experiment, data are collected, but these data by themselves are of no use. They cannot be said to be evidence for a hypothesis apart from some specification of the test to which that hypothesis is being exposed by means of these data. In terms that Mayo adapts from Patrick Suppes, a model of the data is needed, as well as models of the experiment and the hypothesis (Suppes, 1962). Such models are specified in part through the determination of test parameters, and error probabilities are then evaluated based on those models. These error probabilities then allow one to evaluate whether the test employed by experimenters has severely tested the hypothesis.

The referent of the term "the test employed by experimenters" may be anything but clear, however. I show next, by returning to the example of the top quark evidence mentioned at the beginning of this chapter, that in some contexts the referent is quite indeterminate, and that when it is determinate, it may be determined in part by the intentions of experimenters.

The Fermilab Tevatron is a particle accelerator that collides protons with antiprotons at very high energies. CDF searched for the top quark by surrounding these collisions with a complex, barrel-shaped detector designed to measure the properties of particles produced in the collisions and their subsequent decay products. Top quarks, if they existed, would be produced very rarely, in top-antitop ($t\bar{t}$) pairs that would then decay according to certain characteristic patterns, or signatures. Principally, the top quark decays directly to a W boson and a b quark. But these also decay, and CDF's search for the top quark focused on identifying particles whose ancestry could be traced to a top decay. One important signature would involve a high transverse-momentum lepton (either an electron or a muon), three or more high-energy jets of strongly interacting hadrons, and another electron or muon (produced by the decay of a b quark) with low transverse momentum—a soft lepton. The search for events bearing this signature was called soft lepton tagging (SLT).

High momentum, low momentum, and the like are vague terms. CDF sought to make them precise in order to distinguish real top quark decays (signal events) from background processes that might mimic this top quark signature (background events). They did this by choosing the values required of various particle measurements (cuts) to constitute a candidate event. Any collision event that satisfied the cuts would qualify as a candidate event. Having chosen a set of cuts, CDF could then compare the number of candidate events in their data with their estimate of the average number of candidate events they could expect to find in such a data set from background sources alone. The existence of the top quark would manifest itself as a significant excess in the number of candidate events beyond the expected background.

What constitutes a significant excess? Quantitative error statistics can help address this question. CDF had determined for themselves a null hypothesis:

> H: This data sample has been drawn from a population of proton-antiproton collision events that is free of top quark production.

They sought to test this against an alternative hypothesis:

> J: This data sample has been drawn from a population of proton-antiproton collision events that contains some top quark-producing events.[3]

For the purpose of such a test, they defined a test statistic:

> X ≡ the number of candidate events in the present data sample

With these elements in place, they produced a null probability distribution for X. This distribution gives the probability of getting various values for X, assuming that H is true, for the experiment being performed. From the null distribution they could then produce an estimate of the expected background.

After collecting data for 18 months beginning in 1992, CDF had data on approximately 16 million collision events. Among these, they found seven SLT candidate events. Based on their null probability distribution, they estimated that they should expect on average approximately three SLT candidate events from background.[4]

Given that outcome, they then sought to calculate the significance level of their results—that is, the probability of getting seven or more candidate events, assuming the null hypothesis H. They found that, were there no top quark, and were they to repeat their experiment infinitely many times, they would get seven or more candidate events about 4% of the time.

CDF had used the SLT search during a previous data-collecting period in which they found no evidence for the top quark. However, having failed to find the top quark, CDF was able to establish a minimum value for its mass (Abe et al., 1992), since theory dictated that the lower the top quark's mass, the more frequently the particle would be produced, and the more quickly it would show up in their data.

As CDF prepared to begin a new round of data-collection in 1992, some discussed the possibility of changing some of the cuts used in the SLT search. The debate focused on the minimum value required for the momentum of the soft lepton. The minimum value had been set at 2 GeV/c. Some argued that the cut should be moved to 4 GeV/c on the grounds that, given a more massive top quark, leptons with momentum in the range from 2-4 GeV/c were much more likely to come from background than from top quark decays. This argument was not absolutely conclusive, however. The physicists chiefly responsible for the SLT search at the time thought they had good reasons to keep the cut at 2 GeV/c—not least in order to maintain continuity with the earlier search. Two facts about the CDF collaboration at the time of these events set the stage for the controversy that surrounded the SLT analysis. First, CDF members were free to examine new data as it became available. Second, the two physicists who did most of the work on the SLT algorithm worked very independently from the rest of the group.

CDF eventually reported the SLT results with the soft lepton cut at 2 GeV/c. However, some physicists in the collaboration expressed uncertainty regarding both the timing of this choice and the way in which the choice was made. Three of the seven candidate events found by the SLT analysis would be excluded if the cut was moved up to 4 GeV/c, yielding an apparently less-significant result. Some collaboration members consequently worried that the apparent significance of the SLT results was an artifact of a manipulation that created the appearance of a genuine effect out of mere background. Particle physicists consider such manipulation— intentional or otherwise—a sufficiently serious problem to merit a special name: "tuning on the signal."[5] In this case, the availability of data for scrutiny entailed that such manipulation was possible, and the relatively private nature of the SLT development process made it easy for other physicists to worry that it had occurred, even if not deliberately.

To understand what is troubling about tuning on the signal, consider the officially quoted significance level for CDF's SLT search: 0.041. Based on the assumptions CDF was making, if the null hypothesis were true, and one were to repeat infinitely many times an experiment using the same detector, using the same cuts, collecting the same amount of data, and so on, one would get as many as seven candidate events or more only 4.1% of the time.

However, if we know that the cuts used in this case were chosen in such a way as to exaggerate the apparent significance of the results, then we have statistically relevant information about the experimental procedure used to reach these results. Specifically, the procedure followed—including now the procedure for choosing the cuts—has different error characteristics than the procedure on which CDF based their significance estimate of 0.041. That estimate was based on the specification of a reference class, which is a hypothetical population of repetitions of the testing procedure used in the experiment. If experimenters have tuned their cuts on the signal, then a reference class that would otherwise be appropriate would be the wrong choice for calculating that probability.

In short, the problem that the SLT analysis posed for some in the CDF collaboration was that uncertainty over the procedure used for determining the cuts meant uncertainty over the reference class to consider when evaluating error probabilities. For those who regarded the characteristics of the testing procedure as having been rendered unclear, it became difficult to assess reliably the severity of the testing procedure employed with respect to the top quark hypothesis as a whole (see Staley, 2002; 2004).

OBJECTIVE EVIDENCE SUPERVENIENT ON ACTIONS

171

If, as the error statistical theory has it, evidence for a hypothesis requires the passing of a severe test by that hypothesis, then one cannot say whether a particular fact constitutes evidence for a particular hypothesis without reference to some test that the hypothesis in question passes with that result. But even if things like quarks exist independently of our beliefs or actions, it does not seem that tests do. Testing procedures exist through the actions of persons who conduct those tests. The CDF data were not by themselves evidence for the top quark. Such status depends on the testing procedure, which depends not only on the cuts used to define a candidate event, but also potentially on the decision procedure used to determine those cuts. And certainly that cannot exist apart from actions of experimenters.

The example of the SLT controversy points to a potential feedback loop: the fact that a certain outcome would obtain relative to one set of test criteria can stimulate a change in the criteria by the experimenters. Indeed, the adjustment of test criteria to achieve a certain outcome needs to be recognized as itself part of the procedure being enacted in those cases. When the effect of this feedback loop on the error probabilities of the procedure is not taken into account, estimates of severity become biased and misleading.

Furthermore, in a collaboration these test criteria are determined by a collective decision, and different members of the collaboration may have different views of the procedure being enacted, both in terms of explicit data selection criteria and in terms of a broader understanding of the test procedure. CDF members disagreed as to whether the SLT test procedure included a feedback loop. In such cases, it may not be clear just what constitutes the factuality of any specification of the test procedure employed.

The testing of a hypothesis by a collaboration is a collective action, and thus requires some form of shared intention. But collaboration members may intend to test the hypothesis in different ways. If that is the case, there may not be one test that the hypothesis is subjected to. The outcomes of the multiple tests intended by the different members of the group may have divergent evidential statuses. In practice, this problem is avoided largely through the demands of publication. The need to come to agreement on a single document representing the collaboration's collective understanding of their results does much to bring about the requisite shared intention. Individual collaboration members subordinate their intention to test the hypothesis according to particular criteria to their intention to test it according to criteria agreed upon by the collaboration.

In the debate over the SLT analysis, the demands of publication resulted in an agreement on the final choice of transverse momentum cut at 2 GeV/c, in spite of the fact that some collaboration members remained convinced that it was not an optimal choice. (As one collaboration member put it, "I could live with that" [Contreras, 1995].) Absent such agreement, there may be no determinate fact of the matter as to just what test a particular hypothesis is being subjected to, and hence no determinate evidential status for the data.

It cannot be said, then, that CDF's data constitute evidence for the top quark, independently of any person's intentions or beliefs. The testing procedure matters, and it is determined in part by the intentions of the experimenters conducting the test. For some testing procedures, the data may indeed yield evidence that there is a top quark, while for others they may not.

An Achinsteinian Response

Is this compatible with Achinstein's conception of objective evidence? It appears to be. If one accepts the error statistical theory of evidence, one could nonetheless insist that principle β holds for those test-outcome facts that do serve as evidence.

> β_E: If H's passing T with outcome E suffices for E to be evidence for H, then E is a good reason to believe H independently of any epistemic state, including whether or not anyone knows or believes (among other things) that T constitutes the procedure used to test H, or that H's passing T with E suffices for E to be evidence for H.

Achinstein endorses a view similar to this for cases where "selection procedures" for collecting data are evidentially relevant (2001: 213n). The outcome of a testing procedure is person-dependent in the sense that it could not occur unless some persons carried out the procedure in question. Nevertheless, that the procedure was carried out, that it had such an outcome, and that it has specific error probabilities are objective facts in the sense that their factuality is independent of our opinions about their factuality. In the kind of case alluded to earlier, in which the testing procedure is indeterminate, we would have to say that evidential status is likewise indeterminate. Yet when evidential status is, in error statistical terms, determinate, the facts that determine it exist independently of being believed or known, and principle β_E is meant to reflect this point.

However, Achinstein does not accept that E can only be evidence for H relative to a testing procedure (2001: 214). So the second aspect of his response to these issues will be:

> A: Even if there are some cases where we cannot say whether data constitute evidence for a hypothesis without specifying a testing procedure, there are other cases where a particular fact just is a good reason to believe a particular hypothesis, independently of any testing procedure.

173

Whereas on the error statistical theory, evidential status is *always* relative to the testing procedure employed, on Achinstein's account the testing procedure is only sometimes relevant to evidential status.

Proposal β_E is simply an application of principle β to cases where the testing procedure employed is evidentially relevant, and as such *appears* compatible with the error statistical theory. On the other hand, thesis A, if correct, would call the very terms of the error statistical account into question. But I will argue that the kinds of cases invoked on behalf of thesis A can be easily accounted for on the error statistical account, provided we give up β as Achinstein understands it.

The argument for A is straightforward. In some cases, a fact can be seen obviously to constitute evidence for some hypothesis without any consideration of a test procedure. Therefore it is possible for a fact to constitute a reason to believe a hypothesis regardless of the testing procedure used with regard to that hypothesis. (Being seen to be evidence is not required for the evidence to really be such; the point is rather that it is because we do know of such cases that we have reason to accept the philosophical claim that such cases do exist.) To borrow an example from Achinstein, Sam's spots (having certain specified characteristics) are evidence that Sam has measles. This does not strike us as an incomplete statement lacking a truth value prior to the specification of a testing procedure. The statement is true, with or without a specification of the testing procedure employed, if any.

An initial error statistical reply would be that the most that such examples show is that there are cases in which a fact's status as evidence for a hypothesis is somewhat insensitive to just what test is employed among any of the testing procedures we might regard as likely candidates, not that the testing procedure is irrelevant to whether it is evidence. It is hard, though not impossible, to imagine a testing procedure that passes a hypothesis that a patient has measles when spots of this sort are present that fails to meet the requirements of severe testing. In the cases where it is obvious that an evidential relation obtains, without mention of the testing procedure, we typically imagine that a person confronted with a particular observation (of spots on Sam's skin) just responds to that observation with an inference (that Sam has measles). In such a case, the inferential process itself can be regarded as a test. The inference proceeds according to a certain rule or "precept"[6] such as "Whenever I observe spots with such-and-such characteristics, I infer the presence of measles." On the error statistical account the evidential force of the spots requires that inferences drawn according to this precept, regarded as tests of the hypothesis inferred, satisfy the severity requirement. Suppose, however, that one based one's belief that Sam has measles on the spots Sam has, but reached that conclusion by an inference drawn according to the precept, "Whenever I observe spots with such-and-such characteristics, if I make the observation on Monday or Wednesday I infer the presence of measles; otherwise I infer the presence of kidney stones." In an inference drawn according to such a rule, Sam's spots do not function as a good reason to believe that Sam has measles.

174

Achinstein will not accept this line of argument, however. He will say that the argument only shows that a fact E might be evidence for, and hence a good reason to believe, a particular hypothesis H, without it being the case that every person who believes H for the reason that E is justified in doing so. But potential evidence is not relativized to any person or to any epistemic situation. Even though the person reasoning according to this bizarre precept would not be justified in concluding that Sam has measles on the basis of Sam's spots, those spots are nonetheless a good reason to believe that Sam has measles, since they are potential evidence of measles. They would be such even if no one were aware of them.

ABSTRACTION OR IDEALIZATION?

To complete my reply to thesis A, therefore, I must explain the flaw in this concept of "a good reason to believe" employed by Achinstein. In doing so, I will also offer my response to β_E, and in the process suggest a reformulation of β that avoids the difficulties Achinstein's account faces.

Achinstein argues that the principal concern of scientists such as the members of CDF is not simply to demonstrate that their results are *their* evidence for the top quark, or that their results are evidence for the top quark for a person in some particular epistemic situation. Rather, their results are evidence for, and hence a good reason to believe, that the top quark exists—period. What does this mean? How should we understand principle β?

On the one hand Achinstein holds that the reasons involved in potential evidence claims are not limited in their relevance to just some persons or some epistemic situations: "in this sense, if it is reasonable to believe h, it is reasonable for anyone to do so" (Achinstein, 2001: 96). This might suggest that he intends β to be understood in what I will call the "good for everyone" sense:

> β': If E is potential evidence that H, then for any person, regardless of epistemic situation (or any epistemic situation, regardless of its justification) E is a good reason for that person (or any person in that epistemic situation) to believe H.

On the other hand, Achinstein holds that these reasons have their status as reasons quite apart from any beliefs actually or potentially held by anyone, which might suggest a "good, but not for anyone" reading:

> β'': If E is potential evidence that H, then E is a good reason to believe H, but this does not entail anything about whether E would be a good reason for any person (or any person in a particular epistemic situation) to believe H for the reason E (even if in fact it would be).

To reconcile these two notions, Achinstein draws a distinction between an "abstract" and a "non-abstract" sense of "good reason to believe." The correct interpretation of β is then:

> β''': If E is potential evidence that H, then for any person (or any epistemic situation) E is a good reason in the abstract sense for that person (or a person in that epistemic situation) to believe H; but this does not entail anything about whether E would be a good reason in the non-abstract sense for any person (or any person in a particular epistemic situation) to believe H (even if in fact it would be).

To illustrate, let us postulate an experiment just like that performed by CDF and with identical results, but performed by the CDS (Collider Detector at Sfermilab) collaboration. Suppose that the members of CDS came to believe, with good reason yet incorrectly, that there were severe biases in their algorithms, and that consequently the 12 candidate events they identified are not evidence for the top quark (although in fact they are evidence for the top quark). On Achinstein's account, the CDS physicists would *not* be justified, in the non-abstract sense, should they subsequently believe the top quark to exist for the reason that they had found these 12 candidate events. Given their epistemic situation, they should not adopt such a belief.

They *would* be justified in the abstract sense, however, in adopting this belief, because after all those 12 events really are evidence for the top quark.

More generally, according to Achinstein, the aim of scientists in performing experiments is to come to have beliefs about scientific hypotheses that are justified in both senses. That is, they seek to have evidence for their scientific beliefs that justifies those beliefs both in the non-abstract sense (because the evidence constitutes a reason for anyone in their epistemic situation to believe the hypothesis in question) and in the abstract sense (because the evidence constitutes an abstract reason to believe that hypothesis independently of epistemic situation).

Why should we accept this multiplication of categories of reason? There seem to be two sources of justification, one of which is somewhat holistic and philosophical, and one of which is more empirical and historical. The holistic reason, not argued explicitly by Achinstein, is the role that abstract good reasons play in his overall theory of evidence and objective epistemic probability. Assuming that this theory as a whole is preferable to other theories of evidence and probability, and that abstract good reasons are essential to the theory, such reasons escape Occam's razor and we should accept them into our ontology. Although I think the error statistical theory, which does not require abstract good reasons, has advantages over Achinstein's theory, I do not intend to press this line of argument here.

The second reason, which Achinstein does invoke explicitly, is that such abstract good reasons are necessary to make sense of scientific practice. He pursues this argument through the discussion of episodes such as Hertz's and Thomson's work on cathode rays. The following rough sketch is intended to convey some idea of how this argument is supposed to proceed.

In 1883, Heinrich Hertz tested the hypothesis that cathode rays carry an electric charge by passing them through an electric field within a cathode ray tube. No deflection was observed, and Hertz concluded that his results constituted evidence that cathode rays are not electrically charged. J. J. Thomson later suspected that Hertz's experiment was flawed due to residual gas in the tube. He speculated that the charged cathode rays may have ionized the gas in the tube, and that the ionized gases subsequently shielded the cathode rays from the electric field Hertz had introduced. Using superior vacuum technology, Thomson repeated Hertz's experiment in 1897 and observed the deflection that Hertz missed. Thomson then went on to perform further experiments to show that the "corpuscles" constituting cathode rays are much smaller than hydrogen atoms.

Achinstein's interpretation of this episode is that Thomson concluded, not that Hertz's results had been evidence against charged cathode rays but now were not, but that they never were. Although Hertz's results seemed to provide a good reason to believe that cathode rays are not electrically charged, they did not in fact do so (except relative to Hertz's epistemic situation), due to the flaw in Hertz's experiment. Hertz's evidence claim was refuted by Thomson, and replaced by evidence providing a good reason to believe that cathode rays carry a negative charge. As Achinstein puts it, "Thomson could have provided a justification of belief in the electrical neutrality hypothesis simply by citing the results of Hertz's experiments and his own preliminary results. That would have sufficed to justify a belief in the neutrality hypothesis on the part of those physicists (up to 1897) in roughly Hertz's epistemic situation. Thomson wanted to do something more powerful, viz. to provide a good reason to believe that cathode rays are negatively charged—a reason not tied to any actual or hypothetical epistemic situation" (Achinstein, 2001: 34–35). Achinstein concludes that veridical evidence, which requires the existence of abstract good reasons, is in fact the goal of scientific investigation: "A scientist wants to know whether some experimental results reported in *e* provide a good reason for believing a hypothesis *h*—not a good reason for someone in some particular epistemic situation, and not just a good reason for him, but a good reason period, independent of epistemic situations" (Achinstein, 2001: 37).

Achinstein and I agree that Thomson sought reasons for believing in the charged cathode ray hypothesis that would survive improvements in the epistemic situations of his fellow and future physicists—something that Hertz's reasons for believing in the electrical neutrality of cathode rays failed to do. Achinstein believes that he also wanted something more than this, but I do not see the evidence for such a claim.

What is apparent from episodes such as the Hertz/Thomson experiments and the search for the top quark is that scientists in making evidence claims are concerned not to be mistaken—that is, they regard evidence claims as corrigible statements of fact that have their factual status independently of any person's beliefs regarding them. It does not follow that Achinstein's abstract good reasons are the only, or the best, way to make sense of this objectivity.

Are there reasons for resisting the claim that there is an abstract sense of justification that calls for the existence of abstract reasons unrelated to any epistemic situation? I think there are, but I confess that they are merely philosophical. I will try to explain my reservations in what follows.

The claim that some facts constitute reasons to believe certain claims only in the abstract sense serves to resolve the tension between the "good for everyone" and "good, but not for anyone" readings mentioned above. Contrary to the "good, but not for anyone" reading, potential evidence E for hypothesis H constitutes a reason for everyone, regardless of epistemic situation, to believe H. Contrary to the "good for everyone" reading, it does not do this in a sense that entitles us to say that everyone would necessarily be rational in believing H for the reason E. After all, the hypothetical CDS collaboration would not be rational to believe the top quark hypothesis on the basis of their own data, because they believe their own analysis to be so badly biased as to defeat any such evidence claim based on their data.

Achinstein's claim about the aims of science comes to this: Scientists seek evidence not only because they are interested in being rational in the formation of their scientific beliefs, but also because they are interested in (abstract) reasons themselves. This means that, although a scientist seeks to be both abstractly *and* concretely rational, the knowledge of reasons *qua* reasons, quite apart from their role in making one rational in one's beliefs, has independent value.

I doubt whether this claim is true, and I intend to show how one can have a robust conception of the objectivity of scientific evidence without being committed to it.

My own view is that there is a strong connection between the concept of reasons and the concept of rationality. Reasons matter to us because rationality matters. Philosophically, we begin with an idea that there is a difference between arbitrary, unmotivated, or poorly motivated personal beliefs or actions and personal beliefs or actions that are rational—those of which we can say, "Yes, it makes sense to me that a person in that situation should reach such a conclusion." Reasons are brought in to make sense of this apparent difference. Wrenched from these moorings in the rationality of personal beliefs or actions, I find it hard to see of what interest reasons could be.

To sharpen this claim, consider a thought experiment: Suppose a scientist were to be offered a mephistophelean bargain by an evil genius: She would be granted the knowledge of some fact (E) that is evidence for a superstring theory (H), and be made to believe H for the reason E, but only on the condition that no one, not even she herself, would ever be in an epistemic situation such that it would be rational to believe H for the reason that E. I think she should reject this deal. Although she may be interested, as Achinstein has it, in more than a good reason for just some person or just some epistemic situation, it does not follow that she is at all interested in a reason that leaves her just as irrational in her beliefs (in the concrete sense) as if she had adopted them after consulting a ouija board.

I think that Achinstein, on the other hand, is committed to saying that she should at least consider accepting the offer.[7] In coming to know of E, and on the basis of E believing hypothesis H, she can believe H for the reason that E. Even if she is irrational in believing thus in the non-abstract sense (for under the terms of the bargain, she does not even know that E is evidence that H), she will still be justified in the abstract sense of believing H for what is, abstractly, a good reason.

I agree with Achinstein that scientists are not, qua scientists, so self-absorbed that their own rationality is their only concern. The members of the CDF collaboration were intensely interested in the top quark itself, not to mention all of the other exciting physics involved in their experimental pursuits, or the thrill of building one of the most complicated experimental instruments ever devised and making it work. Above all, they wanted to "get it right."[8] What I doubt is whether, in addition to being directly interested in top quarks and bottom quarks and silicon vertex detectors, the elimination of sources of error, and so forth, they were also interested in abstract reasons for the top quark hypothesis. Yes, they were also interested in being correct in their claim that their data contained evidence for the top quark, but the disagreement here is over precisely what that interest amounts to.

I do not pretend to have an argument disproving Achinstein's account of what such interest in "getting it right" amounts to. However, I do intend to propose an alternative that may account just as well for the aspects of scientific practice that Achinstein highlights, that fits with a promising theory of evidence (the error statistical theory), and that retains a single category of reasons tied intrinsically to the idea of rational belief.

I propose that what is needed here is not abstraction, but idealization. I agree with Achinstein that when scientists go in search of evidence they do not merely seek a reason for a particular person or group to believe the hypothesis that evidence supports. Neither would a scientist be satisfied if his evidence merely provided a reason to believe that hypothesis for just some epistemic situation—or rather, for just any epistemic situation. It does not follow that scientists seek reasons that exist apart from any epistemic situation whatsoever.

Instead, I propose that the virtues of Achinstein's account can be retained without the problematic appeal to abstract rationality if we take objective evidence to provide reasons that are relative to ideal epistemic situations. Scientists go to great length to rule out errors in their understanding of their own experiments. This, I propose, is not due to an interest in reasons that make no reference to epistemic situations at all, but instead reflects a concern that future epistemic situations may include knowledge that defeats their evidence claims. I now turn to articulating this proposal.

When an experiment yields evidence E for for a hypothesis H, this evidence does, as Achinstein claims, constitute a reason to believe that H is true. For whom does it constitute such a reason? My proposal is that it does so at least for anyone with a correct understanding of the experimental conditions—anyone in what I will call an "ideal epistemic situation." The next task is to give an account of ideal epistemic situations that is both substantive and defensible.

Experimenters seek to establish evidence claims that will survive any corrections that future inquiry would reveal, even if pursued indefinitely.[9] In "How to Make Our Ideas Clear," C. S. Peirce famously held that the truth regarding any matter could be understood in terms of the long-run consensus, subject to no further corrections, that would ultimately be reached by a community of inquirers who persisted in investigating that matter (Peirce, 1931–58: 5.388–410; see Misak 1991 for a perceptive defense). Although I will not go so far, I think we can employ a similar idea here. At the center of Peirce's approach is the belief that there is an objective fact as to what an extended inquiry into any matter would reveal, if that inquiry were pursued, even if no such inquiry occurs.[10] Suppose that H is a hypothesis, E is an experimental result that has been produced, and T is a testing procedure that resulted in E. I will call S's epistemic state ideal with respect to H, E, and T, just in case S knows everything that an indefinitely extended and detailed investigation would reveal concerning whether E fits H, and whether H's passing T with E is the passing of a severe test. This allows the formulation of a rather different claim about evidence and reasons to believe:

180

γ: If H's passing T with outcome E suffices for E to be evidence for H, then E provides a good reason to believe H for anyone whose epistemic situation is ideal with respect to H, T, and E.

Some points should be noted immediately. First, neither fit nor severity depends on whether H is in fact true, so that the truth of H is not among the relevant facts included in an ideal epistemic state with respect to H, T, and E. Second, this requirement that E provide a good reason to believe H for anyone in an ideal epistemic state is stringent in the sense that it requires *more* than that the evidence claim should survive all actual future inquiry (which might after all end sooner rather than later in some catastrophe). It should be capable of surviving future inquiry as it would develop supposing it to be extended indefinitely. Finally, I am not asserting that the truth of a hypothesis involves nothing more than that no future inquiry would defeat the evidence claims one has made in support of it, nor that the aim of scientists is only for their hypotheses to be empirically adequate. When an experiment generates evidence for a hypothesis, this constitutes a reason to believe that the hypothesis is true, not merely empirically adequate.

Some may balk at my attempt to replace abstract reasons with reasons relativized to ideal epistemic situations, because the latter achieves objectivity of evidence at the expense of requiring the objectivity of counterfactual conditionals. Scientific reasoning, however, is replete with consideration of counter-factual conditionals. In any case, the objectivity of such claims is something that the error statistical theory is committed to already. The commitment implied in γ is simply to the claim that there is a certain class of facts concerning the conditions under which the experiment was performed that are (1) relevant to the evidential judgment at hand and (2) ascertainable in principle, though some may not have been actually ascertained. This class cannot be specified in advance, but will depend on the peculiarities of the particular experiment. It is the job of a good experimenter to try to figure out what they are.[11]

When CDF announced having found evidence for the existence of the top quark, they accompanied their announcement with a detailed argument documenting only part of the exhaustive reasoning and testing employed in justifying their claim. They recounted numerous cross-checks, calibrations, and robustness arguments to establish that their epistemic situation, if not quite ideal, was good enough to make their claim without great fear of being embarrassed by future improvements in their own or others' epistemic situations with regard to their experiment. Skeptics within CDF who worried about biases in the SLT search and elsewhere worried that their epistemic situation was in fact rather far from ideal, and that the collaboration was at significant risk of being embarrassed. Whether satisfied or concerned, the relevant viewpoint seems to have been not the view from nowhere, but the view from possible epistemic situations more complete than their own.

CONCLUSION

Achinstein is committed to thesis A, which is incompatible with the error statistical theory. His defense of A rests ultimately on principle β'''. I do not claim to have refuted Achinstein's β''', much less to have established my own alternative γ. I do, however, claim to have highlighted the conceptual underpinnings of Achinstein's account, bringing into focus its strong dependence on the concept of abstract reasons. In principle γ I have proposed an alternative understanding of the objectivity of evidence claims that does not depend on abstract reasons. I suggest that this alternative proposal does as well as Achinstein's at taking into account the features of scientific practice that motivate his claims, although a full argument for this claim would go beyond the limits of this chapter. Principle γ is compatible with the error statistical theory of evidence. I believe that γ has advantages in terms of ontological parsimony as well as an apparently closer connection to scientific practice, though a full defense of it remains to be articulated.

ACKNOWLEDGMENTS
I am grateful to Peter Achinstein for an illuminating and extended correspondence over the issues discussed here, and for urging me to clarify my own views—not just to state my views

more clearly, but to adopt views that are hopefully clearer than those I started with. I also benefited from helpful discussions with Dianne Brain and with colleagues at Saint Louis University: Susan Brower-Toland, Alicia Finch, Dan Haybron, Scott Ragland, and Joe Salerno.

NOTES

1. Achinstein also defines a stronger concept of veridical evidence that requires, in addition, that there is an explanatory connection between E's being true and H's being true (Achinstein, 2001: 174). His full treatment of potential evidence requires only that, in addition to E and background knowledge B both being true and E not entailing H, there be a high probability, given E and B, that there is an explanatory connection between H and E (Achinstein, 2001: 170).

2. Achinstein accepts that some concepts of evidence are thus relativized, but distinguishes them from the more central unrelativized concepts of potential and veridical evidence. *ES-evidence* is relativized to an epistemic situation. If E is ES-evidence for H, then E is a good reason for anyone in a particular epistemic situation to believe H, whether or not anyone is actually in such an epistemic situation, or, if someone is, whether that person does believe E, H, or that E is evidence for H. *Subjective evidence* is relativized to persons. If E is S's subjective evidence for H, then S believes H for the reason that E. It need not be the case that E is a good reason for believing H, even relative to S's epistemic situation.

3. This constitutes a compound alternative hypothesis, comprising different possible rates of top quark production. That rate (determined by the $t\bar{t}$ production cross section $\sigma_{t\bar{t}}$) is a function of the then-unknown top quark rest mass.

4. The SLT search was just one element in a complicated experiment. CDF's evidence announcement included other sources of evidence not discussed here. Their estimate of the significance of their combined results was 2.6×10^{-3}, which excluded several sources of information regarded as positive cross-checks on their result. See Staley (2004) for more.

5. See Franklin (2002) for a discussion of this type of problem as it arose in several experiments in physics.

6. The term *precept* here is drawn from Peirce. See, for example, his "Theory of Probable Inference" of 1883 (Peirce, 1931–1958: 2.694–754, esp. 735).

7. Presumably, her actual decision would depend on her assessment of the probability that she might become aware of such evidence on her own, thus achieving both concrete and abstract justification. In the case of evidence for supersymmetric strings, this probability should probably be considered quite small. As noted previously, Achinstein holds that scientists want both abstract good reasons and non-abstractly rational beliefs. The question here is whether the former would be of any interest *on their own*. I am troubled, for the sake of both Achinstein's claim and my attempt to argue against it, by the fact that such an outlandish thought experiment seems necessary in order to test our intuitions about how to answer this question.

8. I am stealing this phrase from correspondence with Peter Achinstein.

9. I do not say that this is their only interest. Achinstein claims that scientists seek veridical evidence, which entails that they seek evidence in support of hypotheses that are in fact true. I am inclined to agree with this latter claim, but I wish to separate the question of whether scientists seek true theories from the question of whether they seek reasons to believe those theories that exist apart from any epistemic situation whatsoever.

10. It is therefore important to note that Peirce's account is not offered as an irrealist dissolution of truth. To the contrary, Peirce was offering a pragmatic theory about truth that he regarded as resting upon a quite robust realism.

11. Sometimes even good experimenters fail. When such a failure is pointed out, the good experimenter might say, "If only I'd thought to check that!"

REFERENCES

Abe, F., M. G. Albrow, et al. [CDF] (1992) "Limit on the Top-Quark Mass from Proton-Antiproton Collisions at √s = 1.8 TeV." *Physical Review D* 45: 3921–48.

——. (1994) "Evidence for Top Quark Production in p̄p Collisions at √s = 1.8 TeV." *Physical Review D* 50: 2966–3026.

Achinstein, P. (2001) *The Book of Evidence.* New York, Oxford University Press.

Contreras, M. (1995) Oral History Interview by K. Staley. Tape recording, October 17, 1995. University of Chicago.

Franklin, A. (2002) *Selectivity and Discord: Two Problems of Experiment.* Pittsburgh, University of Pittsburgh Press.

Mayo, D. (1996) *Error and the Growth of Experimental Knowledge.* Chicago, University of Chicago Press.

Misak, C. J. (1991) *Truth and the End of Inquiry: A Peircean Account of Truth.* New York, Oxford University Press.

Peirce, C. S. (1931–1958) *Collected Papers of Charles Sanders Peirce,* 8 vols. C. Hartshorne and P. Weiss, eds. Cambridge, MA, Harvard University Press.

Staley, K. W. (2002) "What Experiment Did We Just Do? Counterfactual Error Statistics and Uncertainties about the Reference Class," *Philosophy of Science* 69: 279–99.

——. (2004) *The Evidence for the Top Quark: Objectivity and Bias in Collaborative Experimentation.* New York, Cambridge University Press.

Suppes, P. (1962) "Models of Data." In E. Nagel, P. Suppes, and A. Tarski, eds., *Logic, Methodology and Philosophy of Science: Proceedings of the 1960 International Congress,* pp. 252–61. Stanford, CA, Stanford University Press.

183

WILL GENOMICS DO MORE FOR
METAPHYSICS THAN LOCKE?

Alex Rosenberg

In his *Notebooks,* Charles Darwin wrote, "Origin of man now solved. He who understands baboon would do more for metaphysics than Locke." Darwin's claim is probably guilty of pardonable exaggeration. After all, he didn't prove the origin of man, and Locke's greatest contributions were to political philosophy, not metaphysics. But it may turn out that Darwin's twenty-first century grandchild, genomics, vindicates this claim with respect to both metaphysics and political philosophy, or at least naturalized versions of them. Here I will focus on the latter claim alone, however. Aside from the special importance that evidence from genomics may turn out to have for hypotheses about the foundations of human institutions of cooperation—hitherto relegated to the status of "Just So Stories"—the revolution in molecular biology shows how technological advance can open up to hard empirical evidence questions that have traditionally been viewed as largely non-scientific. The suddenly broadened prospects for evidential illumination in human affairs that I will illustrate make the questions about confirmation on which Peter Achinstein has focused his philosophical research perhaps more central than they were in the heyday of positivism. If the report on the implications of genomics for human prehistory to follow is at all accurate, the philosophical clarification of how evidence can choose among hypotheses will come to have far greater implications outside of "hard science" than was ever supposed.

From the year that William Hamilton[1] first introduced the concept of inclusive fitness and the mechanism of kin selection, biologists, psychologists, game theorists, philosophers, and others have been adding details to answer the question of how altruism is possible as a biological disposition. We now have a fairly well-articulated story of how we *could have* gotten from nature, red in tooth and claw, to here—an almost universal commitment to morality. That is, there is now a scenario showing how a lineage of organisms selected for maximizing genetic representation in subsequent generations could come eventually to be composed of cooperating creatures. Establishing this bare possibility was an important turning point for biological anthropology, for human sociobiology, and for evolutionary psychology. Prior to Hamilton's breakthrough in the 1960s it was intellectually permissible to write off Darwinism as irrelevant to distinctively human behavior and human institutions. The unchecked contempt with which defenders of the autonomy of the social from the biological operated in their attacks on naturalistic approaches to social processes was both breathtaking and without effective rejoinder.[2]

The major components of the research program—the models and simulations, the comparative ethology—are well known. Once Hamilton showed that inclusive fitness maximization favors the emergence of altruism toward offspring, a virtual riot of ethological activity began to identify previously known cases of offspring care as kin-selected, and to uncover new examples of it. Hamilton was joined by Axelrod[3] in identifying circumstances under which reciprocal altruism between genetically unrelated beings would be selected for, and the community of game theorists began to make common cause with evolutionary biologists in the discovery of games in which the cooperative solution is a Nash equilibrium.[4] This led in turn to the development of models of evolutionary dynamics for iterated games such as cut-the-cake, ultimatum, and hawk versus dove, which show how a disposition toward equal shares, private property, and other norms among genetically unrelated beings may be selected for. An independent line of inquiry at the intersection of psychology and game theory developed an account of emotions suggesting that they too may have been selected for, in order to solve problems of credible commitment and threat in the natural selection of optimal strategies in single games.[5]

But in a sense all this beautiful research remains what Lewontin and Gould once characterized as a "Just So Story."[6] There was no evidence for the hypothesis about how cooperation ever got started or why it persisted, and it seemed unlikely that there ever would be. This is not surprising; after all, behaviors, dispositions, norms, and social institutions are not among the hard parts preserved in the fossil record. There are, of course, comparative ethology, neurophysiology, and neuroanatomy. But at most these provide data from which we can reverse-engineer our way into . . . well, into Just So Stories—hypotheses among which we cannot choose on the basis of independent *evidence*.

There is only one evidential source that stands a chance of doing any better: genomics. Only genomics can still bear hard evidence about human evolution that might test alternative hypotheses about behavior, practices, and events in the distant past—a past that stretches back even before the artifacts on which paleoanthropology relies. In this chapter I explore what evidence genomics can provide about the actual evolution of human cooperation, when combined with phylogeny, comparative ethology, neurophysiology and neuroanatomy, and paleontology. To see its potential, however, let's first consider how it can choose between competing accounts of human prehistory. This will give us an idea of how genomics can turn questions hitherto supposed to be purely speculative into matters open to testable answers.

187

By genomics I mean the comparative and often computational study of the nucleotide sequences and the functional organization of the human genome, and the genomes of many other species of animals, plants, and fungi. The Human Genome Project has given us a first and second draft of the 3 billion base pair DNA sequence of the human genome, and it has so far given us a little more information about the human genome. For instance, it appears that even more of it is "junk" DNA than molecular biologists had thought. This "junk" DNA has no known, or even speculative role in development or normal human function and is just along for the ride, so to speak. And it now appears that there are only about 30,000 to 60,000 genes in our genome, which makes it a small multiple of the size of the fruit fly's genome. But genomics—the comparative study of the DNA sequences of the human and other organisms—will begin to give us the sort of detailed information about our genomes that we never dreamed of, and will give it to us as the result of methods we can automate and turn over to computers. Learning about our genomes and their protein products will cease to require genius, and will at most demand ingenuity. Learning exactly which DNA sequences among the 3 billion nucleotide bases express genes, and which genes they express, is a matter of annotating the DNA sequence the Human Genome Project has provided. Even before the whole sequence came into our hands, comparative genomics was providing evidence of large tracts of history about which only informed speculation had hitherto been possible.

We are inclined to think of history as having begun when written records did, about 3000 years ago in the Near East and a thousand years later in Mesoamerica. But DNA sequence data already in hand can extend our knowledge of the general lines of human history so far back as to turn the Inca empire, the fall of Rome, the building of the Great Wall of China, or the founding of Sumerian Ur into matters of recent history. DNA sequence data can answer derailed perennial questions about human origins and prehistory that have hitherto been the domain of pure speculation. Like the bar code on a can of beans on the supermarket shelf, our DNA sequences are labels from which we can read off date and place of manufacture—not just in geological time, but over the last 200,000 years—with resolving power that already approaches only a few thousand years, just beyond the reach of carbon14 dating. Seeing how fine-grained the resolving power of the genetic bar code is in these cases should give us some confidence it can answer countless other questions hitherto beyond the reach of evidence. But to see this requires knowing a little of the science of DNA sequences.

All mammals inherit their cells' mitochondria and the genes they contain only from their mothers, because the mitochondrion genes aren't in the nucleus of any cell—somatic or germline—and so don't make it into the sperm (which is almost entirely nucleus). But since mitochondria are in every cell, they are in every ovum, and thus in every ovum fertilized by a sperm. By contrast, all males and everyone else who has a Y chromosome inherits it from his (or her) father (the parentheses here accommodate the rare XXY females). Mitochondrial gene DNA (mtDNA) and Y-chromosome DNA can be sequenced. Since individuals differ from one another in gene sequence, it is easy to order a sample of individuals for greater and lesser similarity in DNA sequences—whether in the nucleus or the mitochondrion. The more similar the sequences, the more closely related two people are. Given an ordering of similarity in mtDNA and Y-chromosome DNA among people living today, and comparing it to some mtDNA and Y-chromosome DNA sequence in another species whose age is known, geneticists can work backward to identify an mtDNA or Y-chromosome sequence from which all contemporary sequences must have mutated and descended. In effect, they can draw a family tree of all the main lines of descent among mtDNA or Y-chromosome sequences, and they can date the age of various branches in this family tree. Sequence data for mtDNA were available much earlier than for Y-chromosome data. That led to the conclusion that every human being now living is descended from one particular woman living in eastern Africa—the present Kenya and/or Tanzania—approximately 144,000 years ago. She alone, of all women then alive, has had an unbroken line of female descent from that day to this. Every other woman has had at least one generation of all male descendants, which has caused her mitochondrial sequences to become extinct.

Moreover, the narrowness in sequence variation among extant people reveals that we are ten times more similar to each other in sequence data than, for example, chimpanzees are similar to one another in sequence data. It can also be established that this woman, called Eve by biological anthropologists, lived among a relatively small number of *Homo sapiens* who must have gone through some sort of evolutionary bottleneck, that is, most of our ancestors were killed off at some point in the recent past. As a result there were altogether only about 2000 (± 1000) women alive at the time Eve lived. Subsequent sequencing of a portion of the Y-chromosome has confirmed these conclusions. Indeed, as more and more sequence data come in about single-nucleotide polymorphisms and microsatellite loci, the conclusion has become inescapable (in spite of Chinese reluctance to accept it) that all present *Homo sapiens* are descended from this one African Eve and a relatively small number (about a dozen) of African Adams alive at the same time.[7] This explains why intra-racial differences in gene sequences are larger than interracial ones and why polygenetically (many gene) coded traits have not had sufficient time to assort into separate lineages, which explains why race is not a biologically significant explanatory concept. The genetic similarly among humans suggests further that the obvious visible differences among us in skin color, hair color, facial characteristics, and so forth are of relatively recent origin and are most probably the result of both natural selection in local environments and sexual selection.[8]

Besides telling us where and when we started from, following out differences in more and more available DNA sequences, geneticists have traced the details of early human migration out of east Africa into western and southern Africa, and northward. They have dated the arrival of *Homo sapiens* on each of the continents to within a few thousand years, explaining in some detail the peopling of Micronesia, Melanesia, and Polynesia within the last 6000 years.[9] And beyond chronology, sequence data provides other startlingly detailed revelations about matters of prehistorical narrative that had been thought to be forever beyond answers. For example, consider the question that concerned novelists like Auel and Goulding,[10] and many others: What became of the Neanderthals? Well, Neanderthal DNA is available in bones from the Neander valley in Germany. By comparing mtDNA and the ALU gene sequence—a bit of junk DNA sequence that repeats a distinctive number of times in chimp, *Homo sapiens,* and Neanderthal DNA—it can be shown that these three lines of descent don't share these genes at all, as they would have to if there were any interbreeding among them. This is not surprising in the case of chimpanzees and *Homo sapiens,* of course, but that there was no interbreeding at all between our species and Neanderthal is very significant. That means that *Homo sapiens* either killed off the Neanderthal, gave them all a fatal disease, or otherwise outcompeted them in a common environment. Cro-magnon probably outcompeted them, because there is archeological evidence that both populations existed side by side in Europe over many thousands of years.[11] Similarly, the absence of any non-African Y-chromosome sequences among 12,000 Asian males from 163 different populations shows that the migrants out of Africa replaced any earlier Asian populations, and did not interbreed with them either.[12]

Further research will employ DNA sequence data to uncover the detailed narrative of events we never dreamed of reconstructing, and of other events our non-genetic records have misrepresented to us. For example, consider the origin of agriculture in Europe about 10,000 years ago. How did it happen? There is some archeological evidence to show that farming spread from the Near East northward and westward in Europe. But how? By cultural evolution, one might presume: Farming must have spread as people in one European valley noticed the success of those farming in the next valley to the southeast, and copied their discovery. Others have held that the farmers came out of the Near East, and like Cro-magnon out-competing or extirpating Neanderthal they displaced or decimated local populations, took over their territory, and thus expanded the farming regions. Which hypothesis is right is not a question we could ever have expected to answer, because these events took place before any *recorded* history—indeed, before writing!

But recent studies—first of mtDNA and now of Y-chromosome sequence differences in contemporary near-eastern and European populations—substantiate the latter scenario, the so-called "demic-defusion" model, which is a euphemism for the displacement of one whole population by another. MtDNA and Y-chromosome sequence data show that the earliest migration from the Near East into Europe occurred about 45,000 years ago, and its descendants now account for only about 7% of contemporary European mtDNA. But the earliest immigrants provide twice that proportion of mtDNA among the isolated Basque, Irish, and Norwegian populations, and only half that frequency in Mediterranean populations. The next wave of migration, about 26,000 years ago, provided about 25% of current mtDNA in Europe, whereas the third wave, 15,000 years ago, accounted for about 36% of contemporary European mtDNA. Agriculture arrived with a diffusion from the Middle East about 9000 years ago, and despite their recent arrival the mtDNA sequences these immigrants brought with them account for 23% of the mtDNAs of current European populations—50% when we exclude the extreme Basque, Irish, and Scandinavian populations. And this wave of migration provides mtDNA and Y-chromosome DNA sequences in a *cline*—a gradient of change in proportions—that moves in the direction from southeast to northwest.[13] What the sequence data tell us is that near-eastern populations displaced indigenous ones year after year in a wider and wider arc of expansion from the Middle East, either driving them west, eventually to the extremities of the European continent, or killing them off, so that the only survivors of the original population of Europe were those inhabiting agriculturally marginal territories.

The question arises then, Why didn't the earlier inhabitants acquire farming either independently or by imitation of their neighbors' practices? Surely there is no gene for farming which they lacked. Did farming and the social organization it produced make the near-easterners that much more formidable than the hunter-gatherers? If so, why? Further thought about this displacement should at least enable theorists of the evolution of cooperation among hunter/gatherer egalitarians to set some constraints on their models. The payoffs to cooperation cannot be so strong as to prevent defeat by less egalitarian groups with storable commodities.

191

More recent population events, besides revealing who settled Melanesia, Micronesia, Polynesia, and the Western Hemisphere—and when they did so—will tell us who arrived later, what groups went back the other way to settle Madagascar (where mtDNA sequences are quite different from mainland African mtDNA),[14] and why the current residents of the Andaman Islands east of the Indian mainland have mtDNA sequences far closer to those of east Africans than to those of even the inhabitants of their neighboring islands or the Indian subcontinent.

Non-human DNA sequence data will be able to tell us still more about human prehistory. Sequencing the domesticated plants and animals and their extant undomesticated relatives can tell us where and when hunting and gathering first gave permanent way to farming, and thus to the beginnings of hierarchical social, political, and cultural institutions. And they can date these events well before, or with much greater accuracy, than the archeological evidence now available can. In fact, what DNA sequence research thus far has shown is that both wheat and cattle were probably domesticated at least twice, independently and at roughly the same time. Among the earliest domesticated cereals is emmer wheat, which, however, reflects two different sequences that diverged two million years ago. One is traceable back only to southern and central Europe, including Italy, the Balkans, and Turkey, while the other is ubiquitous to all regions of emmer cultivation. This suggests a double expansion from domestication in the Middle East. There are two distinct types of cattle: the humped breeds of India and the humpless ones of Africa and Europe. They were both domesticated two thousand years after wheat, but their DNA sequences are sufficiently different to support the hypothesis of separate domestication.[15]

Another particularly striking example of an inference from non-human sequence data to human history is provided by Stoneking's conclusion about when humans began to wear clothes. It involves the fact that there are three species of lice: head, body, and pubic lice. Of these, the second is known to be derived from the first, and lives only in clothes. The mtDNA sequence differences between them suggest that the body louse diverged from the head louse about 72,000 years ago, and the greatest diversity in body louse DNA is among those found in Africa. Not only do the data suggest that the niche for body lice—clothing—emerged at this time, but the data also confirm the African origins of *Homo sapiens* as well.[16]

Our question is whether gene sequencing can provide evidence that will distinguish among the alternative hypotheses about how, why, and when human cooperation emerged. If the emergence of cooperation happened in several places independently, then it was presumably a matter of cultural, and not biological evolution. Would it follow that we can immediately excuse genomics from the task of choosing among hypotheses about when, where, why, and how cooperation emerged? Even if gene sequences can shed light on cultural evolution, there are other problems here, that we need to face immediately. First, if the Just So Stories have it right, cooperation, recip-rocal altruism, a sense of justice as equal division, and the emotions and the norms that enforce and express these institutions long antedate agriculture. Moreover, if anything, we should expect agriculture to begin to provide an environment in which many of the dispositions we seek to explain would cease to be selected for. Once agriculture kicks in, the inclination to equal effort and equal shares becomes much less adaptive for individual survival. Storable commodities and capital investment emerge, and the payoffs to cooperating, sharing, reciprocation, defecting, hoarding, and free-riding, not to say domination, become quite different and hard to model, and they probably produce unstable equilibria.

But in fact, the gene sequence data may be relevant to the evolution of cooperation, in spite of all these caveats. The mitochondrial DNA sequences strongly suggest that sometime around or before 144,000 years ago, there was a bottleneck through which *Homo sapiens* came. This was long before the advent of agriculture, and presumably cooperation was already well established at that point (as human paleoanthropological evidence, e.g., cave paintings of hunting groups, shows). If *Homo sapiens* is the sole species in which substantial cooperation emerged, and if we could compare gene sequences between extant and extinct hominid lineages, then there would at least be a chance of uncovering a genetic difference relevant to this pheno-typic difference. There are a lot of ifs here. But even if sociality is a forced move written into our gene sequence and not those of extinct hominids (a tendentious assumption yet to be discussed), it is obvious from gene sequence data that these other hominids left no representatives for us to sequence and compare. Or did they?

Recall that DNA has been extracted from Neanderthal bones upwards of 40,000 years old. This work is part of a new subdivision of biological anthropology that styles itself the study of *ancient DNA*. Quantities of DNA found in burial ground bones, around cave- and camp-fire detritus (including coprolites), in fossil skulls, and so forth are minuscule; proportions of the full sequence are low, and no particular portion—say, functional genes as opposed to junk DNA—is preferentially preserved. Nevertheless, the prospects of acquiring worthwhile data are not entirely unfavorable. The optimism here, as elsewhere in the genetic revolution, is in the power of a molecular process: the polymerase chain reaction (PCR) for the amplification of DNA. This is a process employing a reagent that can catalyze the amplification (reproduction) of a single nucleotide sequence of any length into a million copies, in only thirty rounds of replication. This means that if a molecular biologist can extract just a simple molecule of the DNA from any specimen, an unlimited number of copies will shortly be available for sequencing, comparing to other sequences, and functional annotation (identifying the part of a gene, if any, it codes for). Naturally, the older a specimen the smaller the amount and the shorter the DNA molecules recoverable. Moreover, in sequencing hominid DNA, the greatest stumbling block is contamination with contemporary human DNA, which literally spews from the fingertips of the investigator running the PCR procedure. But (as yet unreplicated) claims of successful amplification and sequencing include 80 million-year-old dinosaur bones, and 130 million-year-old insects trapped in amber.[17] So if (a) sociality is encoded in the genes, and (b) we can find the right specimens, gene sequencing holds out the prospect of answering questions that are otherwise open only to speculation.

But what reason is there to suppose that either of these two assumptions obtains? What indeed would we be looking for, were we to seek genes for cooperation? The first problem is to characterize the phenotype with sufficient precision. Identifying immediate protein products of particular genes is relatively easy, but identifying anatomical phenotypes is no trivial matter. Identifying behavioral ones, assuming they exist, is much trickier. In some cases, however, identifying phenotypes has become much simpler with the advent of positional gene cloning. The strategy involves identifying a trait, usually a disease or deficiency that appears to assort in accordance with Mendelian principles; locating a chromosomal marker in victims that assorts the same way; and then by automated means zeroing in on the specific gene sequence whose mutation or rearrangement is closely correlated with the trait. We thus establish the normal sequence as the gene for the normal trait in the normal environmental range. (It's worth

noting, of course that the expression "gene for" is both widely misunder-
stood and misused. As I use it here, it can at least be quasioperationally
defined as the sequence that would be identified as the wild-type in a
positional cloning exercise. The expression "gene for" must be understood
as always relativized to a population and an environment.) But the oppor-
tunity to employ positional gene cloning will be limited to behaviors that
(in the normal environmental range) are rendered dysfunctional by a single gene error. It is very unlikely that the behaviors we seek, or the dispositions to such behaviors, will be so controlled.

Current inquiry into the genetic basis of behavior begins with the
assumption that behavioral dispositions that are statistically heritable or
disproportionately represented in some genetically homogeneous groups
are matters of degree and are dependent on a large number of genes. For
example, the search for a genetic basis of criminality or intelligence—both
taken to be dispositions measurable by criminal records or performance on
a test—treats the disposition as a quantitative trait and seeks a locus in the
genome statistically associated with that trait in the populations who manifest it in a high degree. These quantitative trait loci (QTL) studies are
both politically and scientifically controversial. Few such studies reveal
even a .20 correlation between the quantitative trait and some region of
the genome on which a detectable marker can be found. QTL studies face
two scientific problems. First, most traits of interest are hard to opera-
tionalize, so that individuals who instantiate them to the greatest degree
are hard to identify. In effect, the traits of interest are not themselves phenotypes, but at most packages of phenotypes or the result of phenotypic
and environmental interaction. Second, QTL studies will at best identify a
set of loci—perhaps 10 or more relatively large stretches of DNA—that are
jointly highly correlated with the instantiation of a high degree of some
quantitative trait in a normal environmental range. Such studies will reveal
nothing about the biosynthetic pathways from these genes to the actual
behavior they are supposed to be the "genes for." It is easy to see how these
problems will bedevil the attempt to employ genomics as evidence to test
alternative theories about how human cooperation emerged.[18]

To make matters concrete, suppose the behavioral disposition we seek to
explain as an evolutionary adaptation is something as specific as the disposition to engage in tit-for-tat strategies in iterated prisoner's dilemma
games; or the disposition to ask for one-half in iterated cut-the-cake
games; or again, the disposition to reject anything less than one-half in an
ultimatum game. Call the first of these dispositions TFT, for short. Now,
no one supposes that any of these dispositions is a single gene-controlled
phenotype such as tongue-rolling. Genes just don't seem likely to code for
recognition of a complex environmental conditional setup in which an

abstractly described strategy is to be employed. Rather, if anything like TFT behavior is actually evinced by humans, then it may simply be the result of a package of genes for much simpler disposition that the TFT behavior supervenes on, without there being a phenotypic TFT disposition to be genetically encoded at all.

Here is a striking example of this sort of thing. The male mouse is disposed to kill all mouse pups not its own offspring—a highly adaptive bit of environmentally conditional behavior that maximizes its genetic representation. But how could nature have programmed the male mouse with the power to make the required genealogical discriminations, given the similarity in look, smell, or other features a mouse can detect in pups? It didn't have to. Instead, nature found a "quick and dirty" substitute that does just about as well. Male mice have a genetically hardwired pup-killer disposition. But mice do not live in large colonies, and nature equips the male mouse with an additional package of genes that automatically switches off the mouse's pup-killer disposition from day 18 to day 22 after its last ejaculation. This period happens to be the gestation time for female mice, so pups the male encounters during this period have a high probability of being its own pups and have a chance to escape before the pup-killer instinct returns.[19] For all the world it looks like male mice show a complicated strategy requiring considerable genealogical knowledge, when in fact the behavior is hard-wired, and the gene that produces it is a quick and dirty solution to a hard problem.

Similarly, TFT behavior will be indistinguishable from behavior generated by some much simpler genetically encoded dispositions. In particular, a gene for unconditional kin-altruism will produce behavior indistinguishable from TFT strategy in iterated prisoner's dilemma circumstances, when all players are close kin. That there is a gene for kin-altruism, or any preferential treatment of kin, or for that matter some quick and dirty substitute for it (a gene for altruism toward anything that secretes a certain odor, for example) among mammals is a pretty safe bet. But if there is a "gene for" kin-altruism or even any quick and dirty available substitute for it, there is also some considerable evidence that such a gene either never figured among the genotype of primates, or that if it did, it made no significant contribution to cooperation among them. This is due to the fact that long before the time of our last common ancestor with the chimps (about 5 million years ago), all the primates had ceased to live in groups in which kin-altruism would be selected for. Or at least that is what a comparative analysis of our closest primate relatives suggests. The social structure of almost all extant ape groups reflects female (and often also male) dispersal at puberty, a high uncertainty of paternity (except for gibbons), an abundance of weak social ties, and a lack of strong ones. Paleontology reveals that the number of ape species underwent a sharp decline about 18 million years ago, while monkey species proliferated. If this were the result of competitive exclusion of apes toward marginal tree-limb niches, it would explain many of the anatomical similarities between them and humans. Unlike humans, chimps and gorillas have remained in these restricted niches to the present. Humans and chimps are highly individualistic, mobile across wide areas, self-reliant, and independent. By contrast, the monkey species reflect matrifocal social networks that would strongly encourage the selection of kin-altruism.[20] At a minimum, the pattern of sociality we and the other primates inherited from our last common ancestor makes it highly probable that cooperation among us is not written in the genes, even imperfectly or approximately, by some quick and dirty exploitation of an already available gene for some form of kin-altruism, and still less by direct natural selection for the disposition to TFT.

All in all, it seems more reasonable to assume that TFT and other cooperative behaviors are the results of the collaboration of a number of different behavioral dispositions, all simply reinforced by their environments, that is, dispositions ontogenetically selected for, though not phylogenetically selected for. If so, it would be worthwhile seeking a package of genes that produce the dispositions and capacities that are individually (non-trivially) necessary, but not jointly sufficient for these sorts of cooperative behavior. (Roughly, a gene is non-trivially necessary for a phenotype if it is not also necessary for a large number of other traits, including respiration, metabolism, reproduction, survival, etc.) In this scenario a great deal of the burden of explaining the exact shape of cooperation is shouldered by the environment in which hominids must have survived for hundreds of thousands of years. And the degree to which our genomes are explanatorily relevant to cooperative dispositions will turn on whether the genes that subserve cooperative behavior were selected for, owing to the fact that they make hitting on TFT or one of the other strategies overwhelmingly likely and make it easy to learn these strategies from others. If they merely make it easy to discover and learn any complex behavior, the notion that cooperation is an evolutionary adaptation naturally selected for will be undercut. The former case, on the contrary, would go some way toward vindicating evolutionary scenarios for cooperation.

Exponents of an evolutionary account of cooperation will favor an account of the matter in which dispositions that specifically subserve cooperation are selected for, owing to the payoff cooperation provides for fitness. Indeed, some will hold that dispositions and capacities useful for other purposes besides fostering sociality—capacities such as memory, speech, and reasoning—have been selected for, owing to their contribution to solving the design problem presented by iterated prisoner's dilemmas and other competitive games. Suppose the genes for a suite of widely useful capacities such as speech, memory, and a theory of mind were all selected for because together they made an agent's seeing and choosing the cooperative strategy a "no-brainer"—an obvious move in appropriate circumstances. We might be tempted to say that together the sequences do constitute a gene for cooperation.

Which of these possibilities obtains is something that gene sequencing may illuminate. We suppose that the genes needed for the evolution of cooperation will include those that subserve general capacities such as memory, reasoning, and speech, as well as those specific to cooperation, such as the emotions of anger, shame, resentment, guilt, love, jealousy, and revenge. One of the ways to begin to identify the relevant phenotypes and genotypes on which cooperative behavior supervenes is to examine hereditary and genetic defects in humans.

For an example of a defect more directly tied to the specific dispositions involved in cooperation, consider that high-function autism and Asperger's syndrome, which prevent normal cooperative behavior, are associated with anatomical and neurological abnormalities in the brain, and (in the case of autism at least) have a substantial hereditary component. There is reason to suppose that autism results from the interactive effects of at least three micro-rearrangements on genes, some of which produce a serotonin transporter. These genes are probably located on chromosomes 7 and 15, and they are implicated in some other rare, genetically caused retardation.[21] We know that normal children develop a theory of mind—the attribution of intentional states to others—between the ages of two and four, and there has been some empirical investigation and a good deal of debate about whether the primates show a similar capacity. If the capacity to treat others as having intentional states is lost in autism,[22] then we are on the way to locating the genes that are either non-trivially necessary, or perhaps even sufficient, for that capacity in humans.

For another example, it has recently been shown that certain significant defects in speech assort in genetically familiar patterns. Positional cloning has enabled geneticists to locate the particular genes responsible for the defect and, *mutatis mutandis,* the genes whose normal function is necessary for normal speech.[23] It occurred almost immediately to the researchers who discovered the "gene for" a hereditary speech disorder that genomic comparisons to chimps could reveal important information about the evolution of language competence, a vital necessity for the emergence of complex cooperative dispositions. We know that chimps and gorillas have shown substantial communicative behavior in domestication, and ethological study of vervet monkeys continues to increase our knowledge of their lexicon well beyond the well-known calls for eagle, leopard, and snake. What these animals appear to lack is syntactic skills, and that these skills are genetically hard-wired in us is suggested not just by Chomsky's speculations, but also by Derrick Bickerton's studies of the transition from Pidgins to Creoles.[24]

The sorts of sequence data available from mitochondrial DNA, or for that matter the Y-chromosome sequence data, are however, completely inadequate to test hypotheses about human/primate genetic differences and similarities. As is well known, to begin with the sequence similarity between *Homo sapiens* and chimpanzees is something over 98%, and the size of the genomes is immensely greater than that of the mitochondria their cells bear. Moreover, approximately 95% of the sequences in both genomes are junk DNA, which does not code for any gene products and whose function, if any, is unknown. Presumably, the differences between *Homo sapiens* and chimps is to be found among the 5% of coding sequences, in the regulatory sequences that control the expression of structural genes identical between humans and chimps. But where these coding sequences are across the 3 billion base pairs—and how they differ—is the issue. Some other source of evidence is needed, and a way of analyzing it.

However, it is at this point that the next generation of genomic data comes into play. For even in the last five years, genomics has moved from comparisons of relatively small amounts of extranuclear DNA to the comparison of entire chromosomes employing automated gene-chip or microarray technology. A gene chip is a small piece of glass on which a huge number of gene sequences can be arrayed. These will preferentially bind to sequences that are closely similar to themselves, when a sample of such sequences are washed over the chip. Those sequences that have bound a similar sequence from the sample can be detected. If the sequences on the chip are known, it is trivial to read off the differences between genes originally placed on the chip and those of the sample. So, once we have located some or all of the genes on, say, a human chromosome, we can array sequences from these genes on a chip, wash the sample with DNA sequences from the homologous chimpanzee chromosome, and read off the sequence differences. If enough is known about how the sequences on the human chromosome realize particular genes, we can identify the presence or absence of the same genes in chimps, as well as differences in their structure, number, and location on the chromosome. If enough is known about the biosynthetic pathways into which these genes enter, we are in position to identify the genetic bases of differences in anatomy and behavior between *Homo sapiens* and chimp.

Such a program of research has already begun to be carried out for the human chromosome 21 and the homologous chimpanzee chromosome 22. Human chromosome 21 is the shortest and was the second to be fully sequenced (by a German-Japanese consortium). It contains only 225 genes (of which 98 are identified only through computer gene prediction) within 33.5 million base pairs. This chromosome is of particular interest owing to its duplication in Trysomy 21 (Down's syndrome) and to the role of genes it carries in Alzheimer's disease, some forms of epilepsy, autoimmune disorders, a form of manic psychosis, and deafness. Starting in March 2003, a microarray comparison between the human chromosome 21 and the homologous chimp chromosome 22 was undertaken.[25] What this work so far shows is (1) that there are not just individual polynucleotide differences, but substantial genomic rearrangements—both insertions and deletions—between the two genomes; (2) that these rearrangements account for about 50% of the total sequence differences between chimps and humans on these chromosomes; and (3) that the deletions at least appear to be random in origin, and both deletions and insertions are randomly distributed across the chromosomes (except for one 250-kilobase region).

199

Let's apply these evidential breakthroughs to the study of the evolution of sociality. Assume that cooperative behavior does not result from a single genetically coded behavioral disposition, but rather that it is taught, learned, and culturally selected for, once it appears. However, there is a suite of hereditary phenotypic dispositions on which it depends. What will these look like? Most skeptics about genetic determinism will claim that these phenotypes are likely to be at most anatomical structures, and are in many cases mediate and immediate protein products of regulatory and structural genes that are at best causally necessary for the behavior but not sufficient for it, even in relatively restricted circumstances. If these skeptics are right, genomics can do little for our inquiry, and not much for human behavioral biology, evolutionary psychology, biological anthropology, or the rest of social science.

But whether they are right or whether there are gene sequences sufficient in normal environmental circumstances for complex behavior is of course an empirical question, and on it some interesting light has already been shed by genomic studies of other model systems. In mice, normal nurturant behavior includes creating nests, cleaning pups, retrieving them to the nest, and crouching over them to provide warmth and nourishment. Nurturance in mice reflects a capacity normally acquired by males and females after exposure to similar retrieval, cleaning, warming and feeding behavior in other mice. When the mouse genome is subjected to a "knock-out mutation" of *FosB*, a gene that codes for a 4.5-kilobase messenger RNA, the result is that mothers ignore their pups and do not gather them, retrieve them, warm them, or feed them, although they do approach and sniff them. This behavior remains unchanged through several pregnancies and in the presence of appropriate modeling behavior by wild-type (normal) maternal mice. Accordingly, we can exclude learning and experience as causes of infant nurturance. Indeed, wild-type mice that have never been pregnant will show nurturant behavior when exposed to newborn pups, whereas *FosB* mutant mice that have never been pregnant show the same non-nurturance defect. Nor is the defect even limited to females: Wild-type males will nurture and *FosB* mutant males will not. When subject to tests for cognitive, olfactory, or hypothalamic related abnormalities (hypothalamus defects are known to influence nurturance), the *FosB* mutant mice show no behavioral deficits or abnormalities. Studies of *FosB* gene expression in normal mice brains have led researchers to conclude that exposure to pups triggers the *FosB* gene in cells of the preoptic area of the hypothalamus to produce a protein that appears to be critical to nurturing. The *FosB* protein is expressed elsewhere in the brain and may have functions in addition to its role in nurturance. However, research has excluded many more basic and non-specific roles for *FosB*—in olfaction, general cognition, perception, and learning—which might lead to defects in nurturance (and other capacities as well).[26]

It is hard to escape the conclusion that this is a "gene for" nurturing in mice. Why should there not be genes for similarly complex behavior in other mammals, up to and including chimps and humans? Unfortunately, the best way to tell whether there are such genes is simply not open to us. The regulations under which both institutional review boards, for human subjects, and animal care committees operate make it unlikely that the protocols under which knock-out and gene-insertion experiments proceed will ever be approved for humans or chimps. Nevertheless, it is worth considering what such experiments could show. Take, for example, the "grammar gene," as Pinker calls it.[27] As described above: many of the affected humans show normal intelligence, "they have trouble identifying basic speech sounds, understanding sentences, judging grammaticality and other language skills" and a genetic market at a locus called the SPCH1 segment of chromosome 7, at a specific regulatory gene *FOXP2*, disrupted in their case by a translocation. The translocation results in the substitution of guanine by adenine in the nucleic acid sequence, and argenine by histidine in the gene product.

Two forbidden experiments immediately suggest themselves. One involves locating the homolog of *FOXP2* in the chimp (it must be there, since it is already known to be expressed in the developing mouse cerebral cortex), and inserting either a normal human *FOXP2* or some portion of it, so that the same regulatory product is produced in the chimp. It is well known that the sorts of regimes already employed to test linguistic competence among chimps reveal a lack of grammaticality—required for complex schemes of cooperative behavior—in their performance. Will the gene-insertion make a difference either to the individual chimp's language learning capacity or to the enhancement of complex communication among chimps? A second experiment, even less permissible, is to locate the homologous gene in chimps and insert it in human infants, then to follow the children's development to determine what sorts of linguistic deficits result.

The same sorts of experiments will repeatedly suggest themselves as positional cloning identifies more and more specific DNA sequences implicated in the blockage of development and the exercise of human capacities and dispositions we suspect are necessary for complex behavior such as TFT (playing tit-for-tat in iterated prisoner's dilemmas). Beyond the limitations imposed on experimentation with human and primate subjects, the real problem with this strategy seems to be the sheer number of genes and gene products that are implicated in these complex dispositions. The gene for nurturance in the mouse is more likely the exception than the rule, among mammals. But even if it is common, the number of gene products involved in complex behavior may well be beyond current computational limits. If upwards of 60% of the coding regions of the genome are devoted to the production of proteins and enzymes expressed in the brain, then even to identify a significant portion of the "genes for" something as complicated as cooperation will be a vast undertaking. But this fact does not detract from the possibility in principle of employing gene sequencing to illuminate the evolution of cooperation.

The gene chip, applied to gene expression in heritable human behavioral deficits, and to chimpanzee brain function, enables us to begin to identify the genes that are necessary for the sort of complex behavior that constitutes social cooperation, and eventually to decide whether they are also sufficient for it in normal environmental circumstances. It will have to be a three-way comparison, including gene expression in the normally functioning brain, hereditarily malfunctioning brains, and chimp brains. Begin by using a microarray to identify the chromosomal locations of gene sequence differences between the normal and the large range of humans with hereditary neurological malfunctions. Given the location on the normal chromosome of this candidate, use the same gene chip method to establish chromosomal locations of the homologous sequences, if any, in chimps. If the sequence is quite similar in size, copy number, relative location, and so forth, assume that it is not among those interestingly necessary for a distinctive human behavioral disposition. If the gene is absent, or different in number, location, introns, and so forth in the chimp, then it is a candidate for being interestingly necessary for distinctive human dispositions.

It will take a very long time to identify all the genes non-trivially necessary for complex cooperative behavior, and to learn what they do: the biosynthetic pathways from them to behavior. But it will not take as long to simply provide a list of locations, alternate sequences, introns, and copy numbers for these genes, without details about their biosynthetic consequences and their macromolecular, anatomical, and ultimately combined behavioral consequences. At this point it should become possible to construct a number of macromolecular scenarios for how linkages, crossover events, mutation, gene duplications and translocations, and other events were selected for, to produce these nucleotide sequences from the common ancestor of humans and chimps. That such genetic alterations that hold the key to our distinctive capacities and dispositions were selected for, or at least were selected because they were carried along by some other gene sequences, is reflected in the differential adaptation of the primate species. Despite the tiny quantitative nucleotide difference between us, the chimps, and the gorillas, the primates are both relatively unsuccessful species, still restricted to a narrow and endangered niche geographically close to the one we started out in—while we bestride the globe. The sequence differences between our ancestors and theirs must have been selected for in the environments we shared.

202

Once the list of locations and sequences for genes without a known function, but nevertheless implicated in distinctively human behavior, is given, the methods employed to date mitochondrial and Y-chromosome sequences can be used to give the order of emergence and perhaps even the ages of these genes. The comparison of human chromosome 21 and chimp chromosome 22 provides evidence that the genetic differences include rearrangements and duplications. Thus there is reason to think that within homologous sequences there will be the single nucleotide polymorphisms—neutral point mutations—that can provide a molecular clock to date the emergence of each of these distinctively human genes before we know much more about them than that they produce a protein that functions in the brain cells. With the right hominid fossils—Neanderthal, and older ones for that matter—and a great deal of good fortune, PCR amplifications can add important data about chronologies or dates of first appearance to our evidential base. (And those sequences, if any, that are entirely missing or that diverge beyond random point mutations may tell us even more, once we have annotated the genes they figure in).

What will the chronology thus established show us? It depends on what the chronology looks like. The alternatives are obvious: Each gene interestingly necessary for distinctively human dispositions emerges at a different date, or all the genes emerge at roughly the same date, or different subsets emerge together. Any one of these outcomes will drive a significant research program in the evolution of human cooperation.

(A) *None of the gene sequences uniquely expressed in human brains are of the same age.*
Then probably each emerged as the solution to a separate design problem, and their joint result, distinctive human cooperative behavior, cannot be mainly attributable to natural selection, but rather to cultural selection. In this case it will be worthwhile annotating these genes roughly in the order of their age. This is done to shed light on stages of hominid cognitive evolution, by determining older genes' roles in cognitive neuroanatomy and behavior and employing reverse engineering to determine what design problem if any each solves, and what design problems it may make it possible for later genes to solve. If ancient DNA from other hominid lines can be recovered in sufficient quantities, then we will learn something about the evolutionary distance between our branch of the hominid tree and other branches, and from the absence in them of some genes that are expressed in our cognitive anatomy and behavior, we may learn perhaps what design problems that they failed to solve led to their extinction.

(B) *The gene sequences can be divided into two or more sets with roughly the same age.*
This result would be extremely significant. It would be empirical evidence for the hypotheses that the genes of equal age were selected for contributing to the solution of the same set of design problems. Accordingly, identifying the function of genes in a package of the same age could provide evidence to test hypotheses about what design problems that humans and/or their hominid ancestors faced and solved.

Assume that, working from chimpanzee DNA and hereditary human defects (employing positional cloning and microarray technology), we have identified and located the group of genes for linguistic communication (or grammaticality), and the group of genes for a theory of mind, and the group responsible for the emotions crucial to commitment in iterated strategic games, and the group of genes for memory and reidentification of fellow-players in competitive and cooperative games—or simply assume that we have identified many of these gene families and some of the genes in each family. If these genes and groups of genes are about equally old, it is reasonable to believe that they were all selected because they were all involved in solving a design problem, or a small number of connected design problems. If we can correlate the concerted emergence of these genes with what we know about environmental changes, paleo-archeology, and demography, the conclusion that they were selected for solving one or a small number of connected problems is further strengthened. I think that such evidence would strongly support the hypothesis that the emergence of human cooperation was, if not a forced move, a neat trick with a Darwinian explanation.[28]

The test of whether I am right in this conclusion turns on two things, of course. First, we need to extract all the data we can from gene sequences, both human and infrahuman. Second, we need to understand better the qualitative and quantitative relations between this data and the hypotheses they can turn from Just So Stories into confirmed or disconfirmed science. This second project is one in which John Locke's successors may be able to make more of an impact than Charles Darwin's. To paraphrase: He who would understand evidence, might do more for the detailed understanding of human origins than Darwin.

NOTES

1. Hamilton, W., "The evolution of altruistic behavior," *American Naturalist,* 97: 354–56.

2. Cf. Sahlins, M., *The Uses and Abuses of Biology,* University of Michigan Press, 1974. The problem of how even to reconcile the theory of natural selection with the possibility of cooperative institutions was so grave that E. O. Wilson insisted that Camus was wrong: It was not suicide that is the only philosophical question, but rather altruism. See E. O. Wilson, *Sociobiology: The New Synthesis,* Cambridge, Harvard University Press, 1976, p. 3.

3. Axelrod, R., *The Evolution of Cooperation,* New York, Basic Books, 1984.

4. Nash, J., "The bargaining problem," *Econometrica,* 18 (1950): 155–62.

5. Frank, R. H., *Passion within Reason: The Strategic Role of Emotions,* New York, Norton, 1988.

6. Lewontin, R., and Gould, S. J., "The spandrels of San Marco and the Panglossian paradigm," *Proceedings of the Royal Society of London,* 205 (1978): 581–98.

7. For an introduction to the African "Eve" hypothesis and supporting data, see Boyd, R., and Silk, J., *How Humans Evolved,* New York, Norton, 2000, pp. 477–83; B. Hedges, "A start for population genomics," *Nature,* 408 (2000): 652–53; and articles there cited, especially Stoneking, M., and Soodyall, G., "Human evolution and the mitochondrial gene," *Current Opinion in Genomics and Development,* 6 (1996): 731–36. For Y-chromosome sequence confirmation and amplification, see Renfrew, C., Foster, P., and Hurles, M., "The past within us," *Nature Genetics,* 36 (2000): 253–54 and papers there cited; Stumpf, M., and Goldstein, D., "Genealogical and evolutionary inference with the human Y-chromosome," *Science,* 291 (2001): 1738–42.

8. For an account of the natural selection of skin colors, see Chaplin, G., and Jablonski, N., "Skin deep," *Scientific American,* 287 (2000): 74–81.

9. See Cann, R. L., "Genetic clues to dispersal in the human populations: retracing the past from the present," *Science,* 291 (2001): 1742–48.

10. Auel, J., *The Clan of the Cave Bear,* New York, Crown, 1980, and many sequels; Goulding, W., *The Inheritors,* London, Harvester, 1963.

11. Boyd, R., and Silk, J., 2000, pp. 484–85; and Gibbons, A., "The riddle of coexistence," *Science,* 291 (2001): 1725–29.

12. Key et al., *Science,* 292 (2001); 1151–53.

13. Richards, M., Mcacaulay, V., et al., "Tracing European flounder lineages in the near eastern mtDNA pool," *American Journal of Human Genetics,* 67 (2000): 1251–76.

14. Gibbons, A., "The peopling of the Pacific," *Science,* 291 (2001): 1735–37, and papers cited therein.

15. Brown, T. A., Allaby, R. G., Sallares, R., and Jones, G., "Ancient DNA in charred wheats: taxonomic identification of mixed and single grains," *Ancient Biomolecules,* 2 (1998), issue 2/3, pp. 184–85; Turner, C. L., Grant, A., Bailey, J. E., Dover, G. A., and Barker, G. W. W., "Patterns of genetic diversity in extant and extinct cattle populations: evidence from sequence analysis of mitochondrial coding regions," *Ancient Biomolecules,* 2 (1998), issue 2/3, pp. 235–50.

16. Kittler, R., Kayser, M., and Stoneking, M., "Molecular evolution of *Pediculus humaanus* and the origin of clothing," *Current Biology,* 13 (2003): 1414–17.

17. Hoss, M. M., Jaruga, T. H., Dizdaroglu, M., and Paabo, S., "DNA damage and DNA sequence retrieval from ancient tissues," *Nucleic Acids Research,* 24 (1996): 1304–7.

18. For an introduction to these QTL studies see Plomin, R., et al., *Behavioral Genetics,* 4th ed., New York, Worth, 2001.

19. G. Perrigo, et al., "A unique timing system prevents male mice from harming their own off-spring," *Animal Behavior,* 39 (1990): 535–39.

20. Maryanski, A. and Turner, J. *The Social Cage,* Palo Alto, Stanford University Press, 1992.

21. Cf. Menold, M. M., Shao, Y., Wolpert, C. M., Donnelly, S. L., Raiford, K. L., Martin, E. R., Ravan, S. A., Abramson, R. K., Wright, H. H., Delong, G. R., Cuccaro, M. L., Pericak-Vance, M. A., and Gilbert, J. R., "Association analysis of chromosome 15 GABAA receptor subunit genes in autistic disorder," *Journal of Neurogenetics,* 15 (2001): 245–59; Herzing, L. B., Cook, E. H. Jr., and Ledbetter, D. H., "Allele-specific expression analysis by RNA-FISH demonstrates preferential maternal expression of UBE3A and imprint maintenance within 15q11-q13 duplications," *Human Molecular Genetics,* 11 (15, 2002): 707–18.

22. Klin, A., Schultz, R., and Cohen, D., "Theory of mind in action: developmental perspectives on social neuroscience," in Baron-Cohen, S., Tager-Flusberg, H., Cohen, D., eds., *Understanding Other Minds: Perspectives From Developmental Neuroscience,* 2nd ed., Oxford, Oxford University Press, 2000, pp. 357–88.

23. Lai, C. S. L., Fisher, S. E., Hurst, J. A., Vargha-Khadem, K., and Monaco, A. P., "A forkhead-domain gene is mutated in a severe speech and language disorder," *Nature* 413 (2001): 519–23 .

24. Bickerton, D., "How protolanguage became language," in Knight, C., Hurford, J. R., and Studdert-Kennedy, M., eds., *The Evolutionary Emergence of Language,* pp. 32–64. Cambridge, Cambridge University Press, 1998.

25. Frazer, K., Chen, X., Hinds, D., Krishna Pant, P. V., Patil, N., and Cox, D., "Genomic DNA insertions and deletions occur frequently between humans and non-human primates," *Genome Research,* 13 (2003): 341–46. See also Locke, D. P., Segraves, R., Carbone, L., Archidiacono, D., Pinkel, D., and Eicler, E., "Large-scale variation among human and great ape genomes determined by array-comparative genomic hybridization," *Genome Research,* 13 (2003): 347–57.

26. Brown, J. B., et al., "A defect in nurturing in mice lacking the immediate early Gene *fosB,*" *Cell,* 86 (1996): 297–309.

27. Pinker, S., "Talk of genes and vice versa," *Nature,* 419 (2001): 465–66, identified by Fisher, S. E., Vargha-Khadem, F., Watkins K. E., Monaco A. P., and Pembrey, M. E., "Localization of a gene implicated in a severe speech and language disorder," *Nature Genetics,* 18 (1998):168–70, et passim.

28. The distinction is first drawn in Dennett, D., *Darwin's Dangerous Idea,* New York, Simon and Schuster, 1995.

11

IS DOMESTIC BREEDING EVIDENCE FOR (OR AGAINST) DARWINIAN EVOLUTION?

Richard A. Richards

On the standard interpretation of Darwin's argument for his theory of evolution, domestic breeding served as evidence for the power of natural selection, by virtue of a causal analogy between artificial and natural selection. But this interpretation of the causal analogy—the causal efficacy interpretation[1]—is highly problematic. In the most important ways, Darwin did not take domestic breeding to provide evidence for the power of natural selection, nor, given what he was in a position to know, would he have been justified in doing so. There is a better way to understand Darwin's use of the causal analogy. He used it heuristically to assist in understanding and extending evolutionary theory. My intentions are to explain what is wrong with this standard causal efficacy interpretation of the causal analogy, and why it is so widely accepted by contemporary scholars. Along the way, I will outline a better understanding of Darwin's argument, and how domestic breeding functions in it.

To make my case, I will be adopting three of the principles of evidence Peter Achinstein presents in his *Book of Evidence*. First, what scientists want in practice is a good reason to believe a hypothesis[2] (Achinstein, 2001: 6–8). Second, evidence is supposed to provide such a reason. Accordingly, a satisfactory concept of evidence must be *strong*. What this means is that for *e* to be evidence for *h*, *e* must provide a good reason to believe *h*. It is not enough to just raise the probability of *h*. The fact that I enter an elevator raises the probability that I will be in an elevator accident, but since that probability is still very low, it is not a good reason to believe the hypothesis that I will be in an elevator accident.[3] Similarly, for domestic breeding to count as evidence for some evolutionary hypothesis, it must give a good reason to believe that hypothesis.

Third, the question "Is domestic breeding evidence for (or against) Darwinian evolution?" can be interpreted in several ways. We can interpret it to be asking whether domestic breeding is evidence for Darwinian evolution independently of what anyone believes or knows, or is in a position to believe or know. In Achinstein's terms, we could be asking whether it is *veridical* or *potential* evidence.[4] Or we might interpret this to be a question about whether domestic breeding is evidence for anyone in a particular epistemic situation—for anyone with certain kinds of knowledge or belief. Achinstein calls this *epistemic situation* (E-S) evidence (Achinstein, 2001: 20). Finally, we might interpret this question to be about the actual beliefs and commitments of particular individuals—whether or not they take domestic breeding as evidence for Darwinian evolution. In other words, we might be concerned about *subjective* evidence. On Achinstein's account, if e is evidence for h in the subjective sense, it is relativized to a particular person or group of persons at a particular time. And in order for e to be evidence for h relative to a person or group, three conditions must be met: (1) that person (or group) must believe that e is (veridical) evidence for h; (2) that person (or group) must believe that h is true or probable; and (3) that person's (or group's) reason for the belief that h is true is that e is true (Achinstein, 2001: 23, 174).

My initial concern in this chapter is with evidence in the subjective sense: What did Darwin believe about domestic breeding and its evidential value for the power of natural selection? In particular, did he believe in the power of natural selection on the basis of a belief in the power of artificial selection? I am also concerned about epistemic situation (E-S) evidence. There are two reasons. First, if we are to understand the historical context and the acceptance of evolutionary theory, we should know whether domestic breeding was evidence for those in that particular epistemic situation. Was Darwin justified in believing what he did, given his epistemic situation? Second, Darwin and his contemporaries had the beliefs and attitudes they had partly because of what they were in the position to believe and know. Unsurprisingly, we often make inferences about subjective evidence—what particular individuals actually believed—partly on the basis of what they were in a position to believe or know. Insufficient attention to Darwin's epistemic situation is, in my view, one reason for the misunderstanding of how domestic breeding functioned in the presentation of his argument.

In "A Puzzle" I outline the standard, *causal efficacy* interpretation of Darwin's causal analogy: the causal efficacy of artificial selection to produce change established a similar causal efficacy of natural selection. On this view, the analogy provided *for Darwin* evidence for the causal efficacy of natural selection. I argue next that, contrary to the causal efficacy interpretation, Darwin (and his contemporaries) thought instead that the analogy, if anything, provided evidence *against* the causal efficacy of natural selection. So domestic breeding was not subjective evidence in this sense. Nor was it evidence relative to Darwin's epistemic situation.[5] In "Varieties of Causal Efficacy" I identify a crucial ambiguity in the term *causal efficacy* that contemporary scholars seem to ignore, and that contributes to the acceptance of the causal efficacy interpretation.

The next section is an outline of a better understanding of Darwin's evidential use of domestic breeding as "experimental." This evidential use, I argue, is not based on the causal analogy. And in "Explaining the Causal Efficacy View" I explain, first, how differences in epistemic situations can lead to mistakes in determining subjective evidence, and how that sort of error might have contributed to the acceptance of the causal efficacy view. Finally, I describe several nonevidential, heuristic functions of the analogy between artificial and natural selection. First, the analogy functioned for Darwin as a way to illustrate the laws of organic nature. Second, it functioned to help in understanding how the principles and processes of evolution work or might work. Third, it served to indicate how evolutionary theory might be extended. Finally, I will indicate how these heuristic functions have been conflated with an evidential function in a recent analysis of the analogy.

A PUZZLE

Throughout Darwin's published and unpublished work, he made repeated references to domestic breeding and the operation of artificial selection. In the *Origin of Species*, for instance, he discussed variation under domestication *before* turning to variation in nature, and in 1868 he published his two-volume *Variation of Animals and Plants Under Domestication*. The standard reading of Darwin's references to domestic breeding is that he argued (and believed) that domestic breeding provided evidence for his theory of evolution through the analogy between natural selection and artificial selection. On this *causal efficacy* reading of the causal analogy, Darwin argued that natural selection and artificial selection were similar in various ways, and because artificial selection was capable of causing great change, we should conclude that natural selection was also capable of causing great change.[6] One advocate of this interpretation, Doren Recker, notes, "The best way to interpret the role of these appeals to domestic cases is to view them as presenting and supporting the analogical goal of this argument, and to view the entire argument as supporting the causal efficacy of natural selection by appealing to the known causal efficacy of artificial selection" (Recker, 1987: 166).

In general, those historians and philosophers who concern themselves with the role of domestic breeding in Darwin's argument seem to hold similar views.[7] Michael White and John Gribben, for instance, in their *Darwin, A Life in Science*, describe Darwin's strategy: "He needed to convince himself, and obtain evidence with which to persuade others, that new species really could be produced by the accumulation of small changes over many generations. He used pigeons to make his case. . . . The variations that Darwin found among the pigeons bred by those fanciers were so great that any zoologist coming across them in the wild would have classified many of them as belonging to different species, some even to different genera" (White and Gribben, 1995: 163). In this passage, White and Gribben suggest that the analogy between artificial and natural selection established—for Darwin—that selection could produce not only great change (from the accumulation of small changes), but also change sufficient for the production of new species.

Whatever the exact formulation of the argument, the causal efficacy view seems to make three assertions. First, Darwin believed that artificial selection was causally efficacious in producing great change. Second, he believed that natural selection was also causally efficacious in producing great change. Third, his belief that artificial selection was causally efficacious was Darwin's reason to believe that natural selection was also causally efficacious in producing great change. In effect, the causal efficacy of artificial selection in producing great change was Darwin's subjective evidence for the causal efficacy of natural selection in producing great change.

This causal efficacy interpretation, however, presents us with a puzzle. The received view in 1859, when Darwin's *Origin* was first published, was that the functioning of artificial selection in domestic breeding actually presented evidence *against* the hypothesis of evolutionary change, and by implication, against the causal efficacy of selection. One argument implying the causal inefficacy of selection was endorsed by Darwin's mentor, Charles Lyell, in the second volume of his 1832 *Principles of Geology*, sent to Darwin on the Beagle voyage (Darwin, 1958: 100). Here Lyell argued that man seems to domesticate the most changeable species. But even in these species there seems to be a universal tendency to revert to original form. This change by selection, therefore, is only temporary. Species must therefore be immutable (Lyell, 1990, vol. 2: 26). The obvious implication of Lyell's analysis is that change by natural selection would also be subject to reversion. We might schematize this "reversion" argument as follows:

The Reversion Argument:
1. Man domesticates the most changeable species.
2. But even in these species there is a universal tendency to revert to original form.
3. Therefore this change (by selection) is only temporary—and therefore limited.
4. If change (by selection) is only temporary, then it is not real (evolutionary) change.

Therefore, species are immutable (by selection).

If natural selection is similar to artificial selection, the obvious implication would be that natural selection—like artificial selection—is unable to produce permanent change, and by implication unable to produce new species, genera, and so on.[8] Here, the causal analogy seems to be evidentially counter-productive for Darwin: Facts about change under domestication were evidence against the causal efficacy of natural selection.

Alfred R. Wallace was well aware of this argument, and in an 1858 paper, jointly published with a paper by Darwin, he began by discussing the analogy and its evolutionary implications: "One of the strongest arguments which have been adduced to prove the original and permanent distinctness of species is, that varieties produced in a state of domesticity are more or less unstable, and often have a tendency, if left to themselves, to return to the normal form of the parent species; and this instability is considered to be a distinctive peculiarity of all varieties" (de Beer, 1958: 268).

So in 1858, 26 years after Lyell's statement of the limits argument, it was no less accepted that change under domestication was temporary. Even evolutionists doubted that artificial selection had ever produced new species (Huxley, 1986: 75). This was the reason Wallace was anxious to defeat the analogical inference, and he continues (emphasis added): "It will be observed that this argument rests entirely on the assumption, that varieties occurring in a state of nature are in all respects analogous to or even identical with those of domestic animals . . . *But it is the object of the present paper to show that this assumption is altogether false*" (de Beer, 1958: 269).

Wallace's concern with this reversion argument is indicative of its widespread acceptance and force. Asa Gray, an American botanist who defended Darwin in an essay from 1860, also discussed the reversion argument, and seemed to accept its premises. He wrote, "It is said that all domestic varieties, if left to run wild, would revert to their aboriginal stock. Probably they would wherever various races of one species were left to commingle. At least the abnormal or exaggerated characteristics induced by high feeding, or high cultivation and prolonged close breeding would promptly disappear; and the surviving stock would soon blend into a homogeneous result . . . which would naturally be taken for the original form; but we could seldom know if it were so" (Gray, 1963: 25).

Significantly, Gray claimed that Darwin's work supports reversion: "Dr. Hooker doubts if there is true reversion in the case of plants. Mr. Darwin's observations rather favor it in the animal kingdom. With mingled races reversion seems well made out in the case of pigeons. The common opinion upon this subject therefore probably has some foundation" (Gray, 1963: 25). And in a later essay from 1875, titled "Duration and Origination of Race and Species," Gray was still concerned that this tendency to reversion not be taken to refute Darwinian evolution (Gray, 1963: 281).

Darwin discusses the reversion argument in his *Origin*. He recognizes that reversion does occur, in particular in domestic pigeons (Darwin, 1964: 25), but he also denies that reversion is universal or inevitable (Darwin, 1964: 15). Then he cryptically suggests (emphasis added), "I may add, that when under nature the conditions of life do change, variations and reversions of character probably do occur; but *natural selection, as will hereafter be explained, will determine how far the new characters thus arising shall be preserved*" (Darwin, 1964: 15). This suggestion that natural selection plays a part in reversion is highly significant, as shall be seen.

There was a second argument implying the inefficacy of natural selection based on the analogy with artificial selection. This argument is based on the well-known fact that the greater the change produced by artificial selection, the greater the difficulty there is in producing additional change. This "limits" argument might be characterized as follows:

> The Limits Argument:
> 1. Man domesticates the most changeable species.
> 2. Modification becomes increasingly difficult the greater the divergence from original form.
> Therefore there are limits to change by selection.

After the publication of the *Origin*, the engineer Fleeming Jenkin endorsed a version of this argument, implying that the analogy with artificial selection is reason to doubt the efficacy of natural selection in producing the necessary kind of change (emphasis added): "The theory rests on the assumption that natural selection can do slowly what man's selection does quickly; it is by showing how much man can do, Darwin hopes to prove how much can be done without him. *But if man's selection cannot double, treble, quadruple, centuple, any special divergence from a parent stock, why should we imagine that natural selection should have this power?*" (Hull, 1973: 306). That Jenkin's objection was taken seriously (see Ruse, 1999: 204) is good reason to worry that the analogy with artificial selection was taken by Darwin and his contemporaries as evidence for the efficacy of natural selection.

What is puzzling is this: If Darwin were making an analogical argument to establish the causal efficacy of natural selection by virtue of its similarity to artificial selection, he would be making an argument that would be regarded by friend and foe alike to have the opposite conclusion—that selection is *not* causally efficacious in the production of new species. In evidential terms: On this causal efficacy interpretation, Darwin believed the facts about change by artificial selection to provide evidence for his theory of evolutionary change by natural selection, and consequently emphasized the analogy to convince others. But others, instead, took these facts about change by artificial selection to be evidence against that theory.

WHAT IS WRONG WITH THE CAUSAL EFFICACY VIEW?

The fact that many of Darwin's contemporaries took the analogy with artificial selection to be evidence against the efficacy of natural selection, does not by itself justify the rejection of the causal efficacy view. After all, Darwin may simply disagree with Wallace and other contemporaries about the evidential value of the analogical argument.[9] Perhaps the analogy was— relative to Darwin—subjective evidence *for* evolution, and relative to Lyell, Wallace, Jenkins, and others, subjective evidence *against* evolution. This response is plausible; people can and do disagree about evidential value. But it is hard to see where Darwin disagreed with Wallace about the analogy between artificial and natural selection. In particular, he agreed with Wallace that artificial and natural selection were very different processes, why they were different, and what the implications were. And most significantly, Darwin, like Wallace, emphasized these differences.[10]

After Wallace rejects the reversion argument in the first few sentences of his 1858 essay, he explains why: Domestic varieties are unlike varieties in nature. "In the domesticated animal all variations have an equal chance of continuance, and those which would decidedly render a wild animal unable to compete with its fellows and continue its existence are at no disadvantage whatever in a state of domesticity. Our quickly fattening pigs, short-legged sheep, pouter pigeons, and poodle dogs could never have come into existence in a state of nature, because the very first step towards such inferior forms would have led to the rapid extinction of the race." (de Beer, 1958: 276). For Wallace, the problem with the artificial selection practiced in domestic breeding is that it produces increasingly unfit breeds, and the greater the selection the less fit the form. Artificial selection is therefore limited by the increasing unfitness of its forms. This is a difference in kind, not in degree. Natural selection favors fitness; artificial selection does not. So with artificial selection we should not expect unlimited, permanent change and the eventual development of new species and genera.

Significantly, Darwin seemed to agree with Wallace about the inefficacy of artificial selection. First, like many others, he had doubts that domestic breeding had ever produced any new species. Huxley reported this in his Darwiniana: "Groups having the morphological character of species— distinct and permanent races in fact—have been so produced over and over again; but there is no positive evidence, at present, that any group of animals has, by variation and selective breeding, given rise to another group which was, even in the least degree, infertile with the first. Mr. Darwin is perfectly aware of this weak point, and brings forward a multitude of ingenious and important arguments to diminish the force of the objection" (Huxley, 1896: 75). While Huxley does not here say what Darwin's "multitude of ingenious and important arguments" involve, it is clear that Darwin agreed with Wallace in distinguishing domestic breeding and nature, and in the reasons for this distinction.

In an 1857 letter to Wallace, Darwin wrote, "I have acted already in accordance with your advice of keeping domestic varieties and those appearing in nature distinct; but I have sometimes doubted of the wisdom of this, and therefore I am glad to be backed by your opinion" (Darwin, 1958: 193). For Darwin, domestic varieties are different from those in nature because of differences in the operation of artificial and natural selection. In an 1859 letter to Wallace, he acknowledged the difference: "You will see what I mean about the part which I believe selection has played with domestic production. It is a very different part, as you suppose, from that played by Natural Selection" (Burkhardt and Smith, 1988, vol. 7: 240).

Like Wallace, Darwin believed that artificial selection was limited because it produced unfit breeds. The reasons for this are clear in an 1859 letter to Lyell: "I daresay selection by man would generally work quicker than Natural Selection; but the important distinction between them is, that man can scarcely select except external and visible characters, and secondly, he selects for his own good; whereas under nature, characters of all kinds are selected exclusively for each creatures own good" (Darwin, 1903, vol. 1: 128). Because artificial selection can be applied only to a limited set of characters and is applied not for the benefit of the breed, but to suit the purposes of the breeder, the domestic breed tends to become increasingly unfit.[11] This can happen relative to an environment. The Niata cattle of La Plata, for instance, have through selective breeding acquired an upturned jaw that prevents their effective browsing on twigs and reeds. In dry seasons they perish at a much greater rate than cattle with normal jaws. This can also happen relative to organization. Short-Faced Tumbler pigeons often have such short beaks they cannot break the shell, and they perish in the egg (Darwin, 1988, vol. 2: 183).

Darwin was keenly aware of this opposition between artificial and natural selection, noting that "When man attempts to breed with some serious defect in structure, or in the mutual relation of several parts, he will partly or completely fail, or encounter much difficulty; he is in fact resisted by a form of Natural Selection" (Darwin, 1988, vol. 2: 183). In these cases, domestic varieties are not "incipient species," because the kind of selection at work in domestication is limited in its efficacy to produce change. The change produced by artificial selection is therefore limited by the increasing unfitness it produces and by the consequential tendency to revert to fitter forms. This is the meaning of Darwin's cryptic comment quoted earlier from the *Origin*: "Natural selection, as will hereafter be explained, will determine how far the new characters thus arising shall be preserved" (Darwin, 1964: 15). In a very real way, natural selection works in opposition to artificial selection—limiting the preservation of characters modified by artificial selection.

What is important here is that Darwin, like Wallace, saw artificial and natural selection as two very different processes—different in kind and not just degree, in that natural selection favors fitness and artificial selection typically reduces fitness. Consequently, Darwin would not have expected artificial selection to produce the kind of changes required. The failure of others to fully recognize the differences between artificial and natural selection caused Darwin in 1866 to lament that he had chosen the term "natural selection," because it connected the two kinds of selection too closely. He wondered if Spencer's "survival of the fittest" might therefore have been a better term (Darwin, 1903, vol. 1: 271).

Varieties of Causal Efficacy

One difficulty in assessing the causal efficacy view is that its formulation in terms of change is ambiguous. We can clarify the issues surrounding causal efficacy by distinguishing the various propositions about change in terms of the following Causal Efficacy Hypotheses (CEH).

CEH1: Natural Selection can produce *directional change*.

CEH2: Natural Selection can produce *great change*.

CEH3: Natural Selection can produce *the kind of change capable of producing new species*.

CEH4: Natural Selection can produce *unlimited divergent change (new genera, phyla, etc.)*.

Directional change is modification over time with a consistent outcome—consistent increase in size, length of beak, elongation of bones, and so forth. *Great change* refers to the amount of directional change. The change producing various breeds of dogs (Great Danes, Pekinese, etc.), for instance, is great directional change. The next proposition, CEH$_3$, is not about the degree of change, but the kind of change—change that favors fitness in various environments, such that natural selection is capable of producing new species. Finally, *unlimited divergent change* is more than just the change that produced the various breeds of dogs, or the change that can produce new species. It is the change that can produce new genera, phyla, and so forth—the full range of existing and extinct life forms.

The first thing to notice here is that these propositions are related—unlimited divergent change requires that the change be of the right kind to at least produce new species—but they are not equivalent to each other. Directional change may not be great change, which may not be the kind of change capable of producing new species or unlimited divergent change (new genera, phyla, etc.). So the fact that natural selection can be shown by the analogy with artificial selection to have the efficacy to produce directional change, or even great change, does not by itself imply that it has the efficacy to produce the kind of change capable of producing new species or unlimited, divergent change.

Second, few in Darwin's time doubted that artificial selection could produce directional change or great change. After all, it had produced pouter pigeons, poodles, short-legged sheep, and exotic flowers. For those who lived in nineteenth-century England, bred dogs and pigeons, and cultivated gardens, this power of artificial selection was obvious. And it was not just farmers that engaged in breeding; many pigeon fanciers came from Darwin's middle class—and above. As Janet Browne explains, "Darwin's Cambridge contemporary Robert Pulleine was a famous exhibitor and poultry judge as well as a vicar, and the majority of men Darwin came into contact with at shows or through his correspondence were profoundly respectable householders from the middling classes, many of them authors of textbooks about bird breeding or contributors to the popular journals which sprang up to cater for their interest. Even Queen Victoria sent her pigeons to competitions" (Browne, 1996: 524).

For Darwin's contemporaries, from farmers to Cambridge scholars, and even the Queen herself, it would already be obvious that selection can produce great change. So while the analogy between artificial and natural selection could certainly provide evidence for Darwin and his contemporaries of great and directional change (efficacy in the terms expressed in CEH$_1$ and CEH$_2$), the evidential value of this is not obvious. If few doubted that selection could produce directional and great change, why would Darwin need to expend effort developing the analogy? He would be providing evidence for what was already widely accepted.

Third, what was really needed by Darwin and Wallace was that the analogy with artificial selection could provide evidence for CEH_3 or CEH_4—that natural selection could produce divergent unlimited change, and the *kind* of change capable of producing new species, genera, and so forth. This is what was really at issue! But the problem is that this is precisely what the analogy could not do. Since the change produced by artificial selection was limited, as Darwin, Wallace, and many of their contemporaries agree, the natural inference would be that the change produced by natural selection was also limited. This is the conclusion that Wallace was so anxious to prevent in the opening statements of his 1858 paper. That is why he argued that artificial and natural selection were really very different, and that is why Darwin seemed to agree. So, for Darwin and his contemporaries, whereas artificial selection could provide evidence for the causal efficacy of natural selection in some weak and trivial sense, it seemed to provide substantial and significant evidence *against* the relevant kinds of causal efficacy.

What this all seems to imply is that, first, the analogy between artificial and natural selection did not provide subjective evidence, relative to Darwin, for the efficacy of natural selection to produce unlimited change, and the right kind of change. He did not believe that artificial selection had that kind of efficacy, and therefore he did not believe that natural selection could have that kind of efficacy on the basis of the similar efficacy of artificial selection. Second, given what Darwin and his contemporaries were in a position to know or believe, the facts established by the operation of artificial selection in domestic breeding did not provide a good reason to believe that natural selection had the causal efficacy to produce the unlimited divergence and the right kind of change. There was no good reason to believe that domestic breeding had produced new species, genera, and so forth, and there were good reasons to believe it could not. Darwin and his contemporaries were therefore not justified in accepting the facts about change by artificial selection as evidence for the causal efficacy of natural selection to produce the change required for evolution. It is my contention here, therefore, that the analogy with artificial selection was neither subjective evidence for the causal efficacy of natural selection in this sense, relative to Darwin, nor evidence relative to his epistemic situation.

DOMESTIC BREEDING AS EXPERIMENTAL EVIDENCE

Even though domestic breeding was neither subjective evidence for Darwin, nor epistemic situation evidence in the manner claimed by the causal efficacy interpretation, it did provide a different kind of evidential support. It provided what Darwin described as "experimental" evidence for, first, the existence and operation of natural selection and, second, the laws of organic nature. This experimental function was important for Darwin because the causal efficacy of natural selection was not the only thing he needed to establish. To appreciate this, we need to recognize that, in the strict sense, natural selection is only survival of the fittest—the differential survival and reproduction of individual organisms on the basis of advantageous traits. Consequently, there were really two questions Darwin needed to answer: First, does natural selection actually exist and operate in nature; is there a survival and reproductive differential on the basis of advantageous traits? Second, what power does natural selection have to produce change?

218

It is important to recognize that these questions are not equivalent. That natural selection actually exists and operates does not establish that it has any particular causal efficacy to produce change. It could exist and operate, and still not produce permanent, unlimited change. There are many reasons why the existence and operation of natural selection do not automatically imply its causal efficacy. For instance, if the advantageous traits are not inherited by the offspring, then there would be no permanent change in the population. There would then be reversion. Or, if there were no new and more divergent variations produced in the following generations, then the change would be limited.[12] Not surprisingly then, Darwin also needed to establish, among other things, that there was a source of new variation, and that favored traits would be inherited—passed on to offspring. He thought domestic breeding was valuable because, among other things, it could provide the inferential basis for the laws governing variation and inheritance.[13]

To fully appreciate what this experimental evidence involves and how it follows from analogical evidence, we need to look at Darwin's *Variation of Animals and Plants Under Domestication*. This work is highly significant because the *Origin*—as Darwin tells us in its introduction—is only an abstract, and does not contain his full reasoning: "This abstract, which I now publish, must necessarily be imperfect. I cannot here give references and authorities for my several statements; and I must trust to the reader reposing some confidence in my accuracy. . . . I can here give only the general conclusions at which I have arrived, with a few facts in illustration. . . . No one can feel more sensible than I do of the necessity of hereafter publishing in detail all the facts, with references, on which my conclusions have been grounded; and I hope in a future work to do this" (Darwin, 1964: 2). The *Variation* is where he presents "all the facts, with references" on which his conclusions were grounded. He makes that clear in the introduction to that later work, where after he summarizes the *Origin*, he tells us in a footnote that he is here stating "the facts on which the conclusions given" in the *Origin* were founded (Darwin, 1988, vol. 2: 1) Part of the reason the causal efficacy interpretation has become so widespread, in my view, is the general neglect of the *Variation*, and the reading of the *Origin* as more than just an abstract. It is easy to get Darwin wrong if we look exclusively to the *Origin* for his argument.

In the *Variation*, Darwin begins by describing domestic breeding as an "experiment": "Man, therefore, may be said to have been trying an experiment on a gigantic scale; and it is an experiment which nature during the long lapse of time has incessantly tried" (Darwin, 1988, vol. 1: 2). The basic idea is that in domestic breeding man has the ability to manipulate various factors and determine the effects.[14] We can, for instance, see how the presence of various artificially produced traits affects the fitness of those who have the traits. *Because* artificial selection produces increasingly unfit forms, and therefore works in opposition to natural selection, it is valuable. According to Darwin, "Natural selection would powerfully affect many of our domestic productions if left unprotected. This is a point of much interest, for we thus learn that differences apparently of very slight importance would certainly determine the survival of a form when forced to struggle for its own existence" (Darwin, 1988, vol. 2: 181–82).

219

When we modify forms by artificial selection, we tend to diminish the fitness of the forms—after all, we are modifying them on the basis of superficial traits for our own purposes, and protecting them from natural selection. By allowing these modified forms to compete with the original, fitter forms, we can determine "experimentally" the existence, operation, and direction of natural selection. Darwin illustrates this "experimental" use of domestic breeding with the Niata cattle: "During ordinary seasons the Niata cattle can graze as well as others, but occasionally, as from 1827 to 1830, the plains of La Plata suffer from long-continued droughts and the pasture is burned up; at such times common cattle and horses perish by the thousands, but many survive by browsing on twigs, reeds, etc.; the Niata cattle cannot so well effect from the upturned jaws and the shape of the lips; consequently, if not attended to, they perish before other cattle" (Darwin, 1988, vol. 2: 183). Here he makes the following observations.

1. In normal seasons, those cattle with upturned jaws (the Niata cattle) graze as well as those with normal jaws. Both groups of cattle survive and reproduce equally well.

2. In drought seasons, because of the change in foliage, the cattle with upturned jaws cannot graze as well as those with normal jaws. The cattle with upturned jaws die at a higher rate than those with normal jaws, and consequently fail to reproduce equally.

In short, because domestic breeding produces unfit breeds we can experimentally determine if a fitness differential results in a differential in survival and reproduction, and the specific reason for that differential. For Darwin this was not a trivial function of domestic breeding. It was experimental evidence for the existence and operation of natural selection—the differential survival and reproduction of organisms on the basis of their traits. It is worth reemphasizing that this experimental evidence for the existence and operation of natural selection did not, for Darwin, establish the causal efficacy of natural selection to produce new species or unlimited change. It merely established a survival and reproductive differential: Those cattle with normal jaws—jaws not modified by artificial selection—survived and reproduced at a higher rate than those with the modified jaws. Notice that this is not an analogical function of domestic breeding. It is not based on the similarities of artificial and natural selection, but on their differences.[15] The very reason that artificial selection is causally inefficacious (it produces unfit breeds) is the reason it can experimentally establish the operation of natural selection.

Darwin also believed that domestic breeding could provide experimental evidence for what he called the "laws of organic nature."[16] We can, he argues in the conclusion to the *Origin*, establish the laws governing heredity, variability, correlation of parts, and so forth.[17] "A grand and almost untrodden field of inquiry will be opened, on the causes and laws of variation, on the correlation of growth, on the effects of use and disuse, on the direct action of external conditions, and so forth. The study of domestic production will rise immensely in value" (Darwin, 1964: 486). Much of the *Variation* is devoted to the inference of these laws, as Darwin makes clear in the introduction: "I shall in this volume treat, as fully as my materials permit, the whole subject of variation under domestication. We may thus hope to obtain some light, little though it may be, on the causes of variability—on the laws which govern it, such as the direct action of climate and food, the effects of uses and disuse, and of correlation of growth. . . . We shall learn something of the laws of inheritance" (Darwin, 1988, vol. 1: 6).

I won't go into the details of Darwin's reasoning here,[18] but the rationale is obvious enough. In domestic breeding, for instance, we can vary the conditions of life for domesticated animals in fairly precise ways, and determine if there is production of new variation of any kind—physical, behavioral, and so forth. From this "experiment" we can, in Darwin's view, infer the causal laws governing variability. We can determine if there is any direct effect from changes in climate or food. Furthermore, we can breed selectively to determine the probability of inheritance of particular traits, or the effect of use and disuse, and we can infer the laws governing these phenomena. For these experimental uses of domestic breeding, Darwin makes the surprising claim that the study of domestic variety is even more important than the study of species in nature: "A new variety raised by man will be far more important and interesting subject for study than one more species added to the infinitude of already recorded species" (Darwin, 1964: 486). Domestic breeding was valuable in Darwin's view because it could provide evidence for evolution by providing the inferential basis of the laws of organic nature. In other words, he thought we could infer the laws of organic nature from the experimental manipulation of plants and animals under domestication.

The bottom line is that for Darwin, even though the facts about change under domestication did not provide evidence for the causal efficacy of natural selection in the most important senses, they did two important things. First, these facts provided evidence for the existence and operation of natural selection—the differential survival and reproduction on the basis of their traits. Organisms modified by artificial selection, like the Niata cattle, survived and reproduced at a reduced rate relative to those not modified, because breeders gave them less advantageous forms. Second, these facts about change under domestication provided the basis for inferring the various laws of organic nature governing important evolutionary phenomena, such as variability and inheritance.

EXPLAINING THE CAUSAL EFFICACY VIEW

Epistemic Situations

If, as I argue, the standard causal efficacy view is mistaken about Darwin's attitude toward the evidential value of the analogy between artificial and natural selection, then we might wonder why this view is so widely accepted by contemporary scholars. Part of the reason, as I suggested above, is the over-reliance on the *Origin* for determining Darwin's argument. In the *Variation* we get not just an abstract, but the details of Darwin's argument. There are also, I believe, other potential explanations to be found—first, in the role of the epistemic situation in inferring subjective evidence, and second, in the conflation of a heuristic with an evidential function.

We can infer subjective evidence—whether something is taken to be evidence for an hypothesis by a particular person or group of persons—from relatively direct information about their views and attitudes. This might include published work, letters and correspondence, notebooks, and so forth. We might also base our inference on somewhat less direct information: what is reported by those who have direct contact with the relevant person or persons. But another, indirect source of information is the epistemic situation of the relevant person or persons: What is that person or group of persons in the position to believe and know, or *not* in the position to believe or know? This use of the epistemic situation makes sense because people have the beliefs they have, partly because they are in a position to have those beliefs. Surely it would be a mistake to attribute a view, attitude, or belief to someone if, given his or her epistemic situation, it was impossible to have that view, attitude, or belief. We could not, for instance, attribute a belief about genetic determinism to Darwin, whose epistemic situation did not contain all the relevant concepts. I am not suggesting, however, that those who advocate the causal efficacy view are making such an obvious mistake, but there are other ways epistemic situations can become problematic in our inferences about subjective evidence.

One problem is that it is often difficult to get a handle on precisely what the relevant epistemic situation involves. We have information about those elements of the epistemic situation that are disputed enough to get discussed in print and correspondence, but there are also elements of the epistemic situation that are not typically discussed. That may be because these elements are regarded as too obvious to be worthy of mention. In this case, the efficacy of artificial selection in producing large, directional change was so well known by Darwin and his contemporaries as not to require argument or discussion. The evidence that artificial selection could produce large, directional change surrounded them—in the flowers of their gardens, the residents of their pigeon coops, the dog sitting at their feet, and the cattle and sheep in the fields of the surrounding countryside. For Darwin, there was little need to argue that selection can produce this kind of change. In his epistemic situation, the power of artificial selection to produce great change was undoubtedly more obvious than it is in ours. This is important, because from the perspective of our epistemic situation, it is easy to overstate the evidential value of change by artificial selection for Darwin and his contemporaries (and anyone in that epistemic situation) in establishing the causal efficacy to produce large change—whether or not it could produce new species, genera, and so forth.

Likewise, the inefficacy of artificial selection in producing new species was obvious for those in Darwin's epistemic situation. Just as they were fully aware of the changes selection could produce, they were also aware of the limits to that change. So when Wallace argued in the 1858 paper against inferring anything about the power of natural selection from the analogy with artificial selection, he was addressing an issue that was an obvious and pressing problem for someone in his epistemic situation. The problem presented by the analogy to Darwin and his contemporaries—that it suggests the causal inefficacy of natural selection—is, on the other hand, far less obvious to someone in our epistemic situation. There are two reasons: First, because we are less intimately acquainted with the mechanics of domestic breeding, we are simply less aware of the limits breeders reach in selection, and the impermanence of these changes. We are only vaguely aware of the reason our highly bred dogs have such short life spans and so many physical problems—the reduction of fitness by artificial selection.

Second, in our epistemic situation, we have reasons to disregard facts that might otherwise count as evidence against its efficacy. In particular, we have reasons to disregard the inefficacy of artificial selection to produce unlimited divergent change. A thorough analysis of these disregarding factors[19] is beyond the scope of this chapter, but we can see what reasons might function in this way.

Most obviously, we have evidence not available in Darwin's epistemic situation, evidence that allows us to disregard the inefficacy of artificial selection. This is evidence that favors the causal efficacy of natural selection on other grounds, and allows us to dismiss what otherwise might count as evidence against the causal efficacy to produce new species and genera. One line of evidence for the causal efficacy of natural selection is found in what we now know about evolutionary change in microorganisms. Because bacteria reproduce and change at relatively high rates, significant change by natural selection can be observed in the way bacteria develop resistance to the drugs used to treat the diseases they cause.[20] New strains are now known to appear on a relatively regular basis. Similarly, many organisms have now been observed to develop resistance to pesticides by natural selection.

A second line of evidence for the causal efficacy of natural selection is found in mimicry, first described in 1862, and over time becoming increasingly apparent to evolutionary biologists. Ernst Mayr explains, "The first proof of natural selection was the discovery of mimicry. The tropical explorer Henry Walter Bates (1862) observed in Amazonia that some palatable species of butterflies had the same pattern and coloration as sympatric toxic or at least unpalatable species, and that wherever the noxious models varied geographically, the palatable mimics followed the same geographic variation" (Mayr, 2001: 122).

That one species can change over time to resemble another to the degree observed is now taken to show that natural selection can produce large directional change of the right kind. Similarly, co-evolution is now regarded as evidence for the power of natural selection. In co-evolution, different species evolve together for the benefit of each species. Pollinators and flowers co-evolve, each adapting to the other. Parasites become adapted to their hosts, and vice versa. It is now possible to see the same phylogenetic patterns in the parasites as in the hosts, as they evolve, speciate, and diverge together (Mayr, 2001: 211). These patterns of change are all now seen as evidence that natural selection has, and can produce the necessary kinds and degrees of change.

A third line of evidence for the causal efficacy of natural selection is found in recently discovered evidence for speciation. One striking example is the apparent evolutionary "explosion" of cichlid fish species in Lake Victoria. It is now believed, on the basis of genetic data, that in the 14,000 years since Lake Victoria refilled after drying up, a single species of cichlid entered the lake from one of the feeder streams, spread to all of the various underwater habitats, and developed into 500 strikingly different species (Zimmer, 2001: 89–91).[21] "The cichlids that live in Lake Victoria today are all close relatives of one another, and only distantly related to cichlids in other lakes and rivers. . . . Their genes show that a single lineage of mouth-brooder came to the lake after it refilled, and then, in the time that it took for humans to build civilization, 500 species were born. Look into the waters of Lake Victoria with an understanding of evolution, and you see a biological explosion" (Zimmer: 90).

There may well be other factors that allow us to disregard the inefficacy of artificial selection,[22] but it should already be clear that in our epistemic situation, we have the luxury of being able to ignore the evidential disvalue of domestic breeding, because we have much more evidence from nature, perhaps overwhelming evidence, that natural selection is efficacious in the most relevant ways. It is not that this new evidence causes us (in our epistemic situation) to reinterpret the facts about change under domestication. It simply gives us a reason to ignore the evidential disvalue of these facts. After all, if we have what we take to be conclusive evidence for the causal efficacy of natural selection on other grounds, we can simply not take seriously what might otherwise count as evidence against its causal inefficacy.[23] Given that we can do this, it would be easy for us to miss the significance of this disvalue in inferring Darwin's attitude about the evidential value of domestic breeding. In short, because the analogy between artificial and natural selection does not have the same evidential disvalue in our epistemic situation, we miss it in Darwin's.

EXPLAINING THE CAUSAL EFFICACY VIEW
Heuristic Functions

There is yet another reason why contemporary commentators might have mistakenly attributed the causal efficacy view to Darwin. He did indeed employ analogies between domestication and nature, but he used them heuristically, and in a variety of ways. These heuristic functions were intended to assist in understanding how evolutionary principles and processes work or might work, how these principles and concepts are related, and how evolutionary theory might be extended.

The first heuristic function for domestic breeding, in Darwin's presentation of his argument, is as illustration. Here it is not by analogy per se that domestic breeding functions, but on the grounds that domestic breeds and species in nature operate *identically* in accordance with the laws of organic nature. Darwin could therefore give examples from domestic breeding to illustrate the operation of these laws. In his *Origin*, for instance, he discusses the law of correlation of growth, then cites cats, dogs, and pigeons as examples: "There are many laws regulating variation, some few of which can be dimly seen, and will be hereafter briefly mentioned. I will here only allude to what may be called correlation of growth. . . . Breeders believe that long limbs are almost always accompanied by an elongated head. Some instances of correlation are quite whimsical: thus cats with blue eyes are invariably deaf . . . hairless dogs have imperfect teeth . . . pigeons with feathered feet have skin between their outer toes; pigeons with short beaks have small feet, and those with long beaks large feet" (Darwin, 1964: 11–12).

The value of this heuristic use should be obvious. Darwin's contemporaries would have been much more familiar with the variation, heredity, correlation of growth, and so forth in domestication than in the wild, because they lived with their domesticated creatures and observed and acted on these principles in breeding them. Wallace was aware of this heuristic function in Darwin's presentation of the argument. So even though he forcefully rejected the view that artificial and natural selection are equivalent in his *Darwinism*, he seemed to approve of the heuristic use of domestic breeding as illustration: "I have endeavoured by means of a series of diagrams, to exhibit to the eye the actual variations as they are found to exist in a sufficient number of species. . . . It will be found that, throughout the work, I have frequently to appeal to these diagrams and *the facts they illustrate, just as Darwin was accustomed to appeal to the facts of variation among dogs and pigeons*" (Wallace, 1989: vi, emphasis added). It should be recognized that the facts here can also function evidentially. A single case of inheritance can both illustrate a law and, along with other cases of inheritance, provide evidential support for that law. This is because the operation of these laws under domestication is not analogical, but identical to their operation in nature.

225

The situation is different, however, with the heuristic analogies. The analogy with domestic breeding functioned heuristically for Darwin, first as a way of facilitating the understanding of various processes, principles, and concepts, and second as aiding in the development of evolutionary theory. Darwin's strategy in both of these functions was, first, to explain how a process or principle works in domestication, and then to explain how an analogous process or principle operates—or might operate—in nature. Then, typically, he would look for evidence in nature.[24] The first type of heuristic functioning—facilitating the understanding—is most apparent in Darwin's attempt to show how selection might be able to adapt. One of the difficulties that he encountered was the inability to imagine how, or understand how, a selection process can produce adaptation. In a letter to Hooker, for instance, he offers to write to W. H. Harvey, a botanist, and "try to make clear from analogy with domestic production the part which I believe selection has played" in adaptation.

One way Darwin did this was to explain in his Essay of 1844, and then in an 1857 letter to Asa Gray, how an imaginary "super-breeder" might produce "adaptation": "Now suppose there was a being, who did not judge by mere external appearance, but could study the whole internal organization—who was never capricious—who should go on selecting for one end during millions of generations, who will say what he might not effect!" (de Beer, 1958: 264).

This imaginary analogy was based on a real one from domestic breeding, and Darwin also turned to that to facilitate understanding. He believed it was easy to understand how domestic breeding could result in a form of adaptation: "On the view here given of the all-important part which selection by man has played, it becomes at once obvious, how it is that our domestic races show adaptation in their structure or in their habits to man's wants or fancies" (Darwin, 1964: 38). Then that which was easy to understand about domestic breeding could promote understanding of an analogous process in nature: "If then we have under nature variability and a powerful agent always ready to act and select, why should we doubt that variations in any way useful to beings, under their excessively complex relations of life, would be preserved, accumulated and inherited? Why, if man can by patience select variations most useful to himself, should nature fail in selecting variations useful, under changing conditions of life, to her living products?" (Darwin, 1964: 469).

The structure of the analogy here can be seen in the following schematization:

226

Domestic Breeding	Nature
1. Variation is present.	1. Variation is present.
2. Some traits are useful to breeders.	2. Some traits are useful to organisms.
3. There is a selective agent: the Breeder.	3. There is a selective agent: Natural Selection.
4. The agent engages in consistent long-term selective breeding for favored trait.	4. There is consistent long-term differential survival and fitness based on the possession of a trait.
5. The favored trait is inherited by offspring.	5. The favored trait is inherited by offspring.
6. Over time the organism becomes adapted to man's needs.	6. Over time the organism becomes adapted to satisfy its needs.

For Darwin, the relations between the ideas and processes on the left side were helpful in understanding the relations between the ideas and processes on the right side. More particularly, that adaptation in one sense can be produced by the processes on the left side, was instructive in understanding how analogous processes on the right side can produce adaptation in another sense. Notice that some of the terms on the right side are identical to the corresponding term on the left—the presence of variation and the laws of heredity are identical in each analogue. But notice also that some of the corresponding terms and properties are not identical, and are in fact different in highly significant ways. First, for whom the traits are advantageous, and therefore what makes a trait good or bad, is different in the analogues. In domestication, favored traits are good for the breeder; in nature, favored traits are good for the organism. Second, the nature of the selective agent is different. The breeder selects on the basis of a desired end, and selects only on what is visible. Nature selects without a conscious end, and on all traits, visible or not.[25]

It is noteworthy that after the passage quoted most recently above, Darwin dropped the analogy, in that he emphasized the differences rather than the similarities between artificial and natural selection—the scrutiny of the whole constitution by natural selection, and favoring the good of the creature rather than the good of the breeder: "What limit can be put to this power, acting during long ages and *rigidly scrutinizing the whole constitution, structure, and habits of each creature, —favoring the good and rejecting the bad?* I can see no limit to this power, in slowly and beautifully adapting each form to the most complex relations of life. The theory of natural selection, even if we looked no further than this seems to me to be in itself probable" (Darwin, 1964: 469, *emphasis added*). The significance here is that it is precisely these differences that Darwin believed make artificial selection limited and natural selection unlimited. *It is how natural selection differs from artificial selection that he believed made his theory probable!* Darwin seems to be emphasizing these differences, to prevent the limits of artificial selection from being taken as evidence for the limits of natural selection. In effect, Darwin is trying to use the analogy heuristically, but at the same time prevent it from functioning as evidence against his theory of natural selection, by emphasizing the differences.

The second way the analogy functioned heuristically for Darwin is in the development and extension of the theory. Not only can an analogy help understand how processes might work, it can suggest ways to develop the theory. One problem Darwin faced was in explaining how natural selection might do more than just adapt, by producing divergent change.[26] He uses the analogy to address this problem: "As has always been my practice, let us seek light on this head from our domestic productions. We shall here find something analogous. A fancier is struck by a pigeon having a slightly shorter beak; another fancier is struck by a pigeon having a rather longer beak; and on the acknowledged principle that 'fanciers do not and will not admire a medium standard, but like extremes,' they go on (as has actually occurred with tumbler pigeons) choosing and breeding from birds with longer and longer beaks, or with shorter and shorter beaks" (Darwin, 1964: 111–12). Darwin refers to this as a "principle of divergence": "Here, then, we see in man's productions the action of what may be called the principle of divergence, causing differences, at first barely appreciable, steadily to increase, and the breeds to diverge in character both from each other and from their common parent" (Darwin, 1964: 112).

Darwin then asks if such a principle might operate in nature, and gives reasons from nature why it would: *"But how, it may be asked, can any analogous principle apply in nature?* I believe it can and does apply most efficiently, from the simple circumstance that the more diversified the descendents from any one species become in structure, constitution, and habits, by so much will they be better enabled to seize on the many and widely diversified places in the polity of nature, and so be enabled to increase in number" (Darwin, 1964: 112). The basic idea is that the greater the similarity between two individuals, the greater the competition for resources. Two nearly identical individuals will be competing in the same niche, and for the same resources. However, if there are differences in form and habits, then there will also likely be differences in the resources pursued, and that will result in less severe competition between these individuals. In short, divergence is favored because it allows the pursuit of a greater range of resources.

We can schematize this analogy as follows.

Domestication	Nature
1. Variation is present.	1. Variation is present.
2. There is a selective agent: the Breeder.	2. There is a selective agent: Natural Selection.
3. The agent notices the extremes in variation.	3. The extremes in variation tend to occupy distinct niches.
4. There tends to be a preference for the extremes.	4. Natural Selection favors the extremes because there is less competition.
5. Over time, and with consistent selection, modification favors the extremes.	5. Over time, and with consistent selection, modification favors the extremes.

Notice that, like the previous analogy, some of the properties and processes are identical, and some are not. Fitness is favored in nature, but not under artificial selection. Here again, it is not an "identical property" analogy. When Darwin presents the analogy he seems to be interested in using the analogy to provide ideas for *how* divergence might work, not to give evidence for the principle of divergence in nature. Shortly after the passages quoted above, Darwin gives his evidence: "It has been experimentally proved, that if a plot of ground be sown with one species of grass, and a similar plot be sown with several distinct genera of grasses, a greater number of plants and a greater weight of dry herbage can be raised" (Darwin,

1964: 112). Notice that this is evidence from nature, not from the analogy with artificial selection. Whether or not it might be possible to infer, from the analogy with artificial selection, that natural selection results in divergence, Darwin did not seem to do so. He looked for evidence in nature. Here again, Darwin is using the analogy heuristically, but not evidentially.

What is striking about Darwin's presentation of the causal analogy between artificial and natural selection in both of these heuristic uses, is that after emphasizing some of the similarities to facilitate understanding or extend his theory, he typically turns to the differences, and expends great effort in emphasizing their significance. There is a long passage in his chapter on Natural Selection in the *Origin* that is particularly instructive and worth a closer look. Notice that it begins with a question, but then highlights the extent and significance of the differences—the disanalogy—between artificial and natural selection.

> As man can produce and certainly has produced a great result by his methodical and unconscious means of selection, what may not nature effect? Man can act only on external and visible characters: nature cares nothing for appearances, except in so far as they may be useful to any being. She can act on every internal organ, on every shade of constitutional difference, on the whole machinery of life. Man selects only for his own good; Nature only for that of the being which she tends. Every selected character is fully exercised by her; and the being is placed under well-suited conditions of life. Man keeps the natives of many climates in the same country; he seldom exercises each selected character in some peculiar and fitting manner; he feeds a long and a short-beaked pigeon on the same food; he does not exercise a long-backed or long-legged quadruped in any peculiar manner; he exposes sheep with long and short wool to the same climate. He does not allow the most vigorous males to struggle for the females. He does not rigidly destroy all inferior animals, but protects during each varying season, as far as lies in his power, all his productions. He often begins his selection by some half-monstrous form; or at least by some modification prominent enough to catch his eye, or to be plainly useful to him. Under nature, the slightest difference of structure or constitution may well turn the nicely-balanced scale in the struggle for life, and so be preserved. How fleeting are the wishes and efforts of man! How short his time! And consequently how poor will his products be, compared with those accumulated by nature during whole geological periods. Can we wonder, then, that nature's productions should be far "truer" in character than man's productions; that they should be infinitely better adapted to the most complex conditions of life, and should plainly bear the stamp of far higher workmanship? (Darwin, 1964: 83–84)

This is not all. After this passage, Darwin continued for nearly a full page reemphasizing the differences between artificial and natural selection. He especially focused on the fact that natural selection "scrutinizes" all characters and selects only for the good of the organisms, whereas artificial selection scrutinizes only a small portion of characters and does not select for the good of the organism. What we need to pay special attention to is that, first, Darwin only asked what natural selection may accomplish. He did not conclude from the analogy what it in fact *could* accomplish. Second, he highlighted precisely those differences between artificial and natural selection that would prevent artificial selection from being evidence *against* the causal efficacy of natural selection to produce the important kinds of evolutionary change—change favoring fitness. If Darwin were making an analogical argument, why would he be so circumspect about the power of natural selection, and why would he spend so much effort emphasizing these differences? We would instead expect him to highlight the similarities, and either downplay or ignore the differences.[27] It is my contention that Darwin was using the analogy heuristically, then trying his best to prevent it from functioning evidentially against his theory, by emphasizing and reemphasizing the differences between artificial and natural selection.

230

By not taking seriously the problem posed by the recognized inefficacy of artificial selection, and by not carefully distinguishing the different kinds of causal efficacy, and the heuristic uses from evidential uses, it is easy to misunderstand what Darwin is doing. A conflation of the heuristic with an evidential function of the analogy between artificial and natural selection is apparent in a recent analysis of Darwin's presentation of his argument. Susan Sterrett argues that Darwin distinguishes methodological and unconscious selection,[28] and that the former is intended to be analogous to divergence in nature, and the latter is intended to be analogous to extinction. She argues that these references are intended to establish the causal efficacy of natural selection[29] (Sterrett, 2002: 152). She is surely right in that Darwin distinguished methodological and unconscious selection, and that he drew analogies here, but she does not carefully distinguish the heuristic and evidential uses. In her second paragraph she asks us to consider the analogy and then describes it as heuristic device, but she concludes with a claim about how it supports a causal efficacy inference.

But exactly how does the analogy go? The analogy drawn between what Nature has done and what humans have done is supposed to help us *understand how* species originated. It is based on comparing the diverse species the reader knows to exist in plants and animals in nature and the diverse varieties the reader has just been told can be produced in domesticated plants and animals. Darwin claims that there are analogues of the principles of artificial selection in nature, then *shows how* those principles produce diverse varieties in domesticated animals and diverse varieties and species in nature. What I will show in this paper is that it is the analogy between two different principles familiar from his studies in artificial selection and the two different principles he claims are operative in nature that provides the main structure and force of the analogy he uses *to make his case for the power of natural selection to produce new species.* (Sterrett, 2002: 152, emphases added)

Hopefully, from my earlier discussion, it is now apparent that Sterrett is referring to different and independent functions of the analogy—the "understand how" and "show how" are heuristic functions and do not automatically support the evidential goal—making the case for the power of natural selection to produce new species. It should also be apparent by now that the causal efficacy to produce change, even divergent change, is not sufficient to establish, for Darwin and his contemporaries, the power of natural selection to produce new species, let alone new genera, phyla, and so on.

The problem with Sterrett's conclusion about the causal efficacy of natural selection to produce new species is that the passages she cites seem to be heuristic in the ways I have outlined above, and where Darwin only asks about selection in nature. For instance, "Darwin motivates his account of the effects of natural selection by posing a question: 'Can the principle of selection, which we have seen is so potent in the hands of man, apply in nature? I think we shall see that it can act most effectually'" (Sterrett, 2002: 156). And several pages later, Sterrett quotes from the long passage I just cited, where Darwin asks whether there might be a process in nature similar to artificial selection: "Time and again Darwin compares Nature as an agent to 'Man' as an agent; for example: 'As man can produce and certainly has produced a great result by his methodical and unconscious means of selection, what may not nature effect?'" (Sterrett, 2002: 158). But then she neglects to explain the rest of the passage, where Darwin lists at great length the differences between artificial and natural selection. Darwin's focus on the differences between artificial and natural selection is inexplicable if we do not distinguish the different uses of the causal analogy between artificial and natural selection.

CONCLUSION

By now it should be apparent that the answer to the question in the title of this chapter is not likely to be a simple one. Domestic breeding was evidence both for and against Darwinian evolution. Relative to Darwin, it was experimental evidence for the existence and operation of natural selection and the laws of organic nature. Relative to many of his contemporaries, it was evidence against the causal efficacy of natural selection to produce new species, genera, and unlimited divergent change. Since both Darwin and Russell believed that artificial selection was not efficacious in this way, and should not be expected to be, they did not want it to be regarded here as providing evidence at all. Darwin's use of the analogy between domestication and nature was instead heuristic. In this function it could help in understanding the various evolutionary processes and principles, and help in extending and developing evolutionary theory.

In order to keep clear how domestic breeding functioned in Darwin's argument, we first need to distinguish the different kinds of functioning—experimental, evidential, and heuristic. We also need to recognize how Darwin would be expected to support the various functions, and how he would be expected to prevent other functions. Then it becomes clear why he typically followed a discussion of the analogy between artificial and natural selection with an explanation and emphasis of the disanalogy. Second, we need to be conscious not just of how Darwin viewed domestic breeding, but how his contemporaries viewed it—as providing evidence against unlimited change. Surely Darwin was well aware of their views, and presented his theory accordingly. This is clear in his constant emphasis of the disanalogy between artificial and natural selection—just as Wallace emphasized the disanalogy. Third, we need to clearly distinguish Darwin's epistemic situation from our own. Even though he disagreed with many of his contemporaries about the evidential disvalue of domestic breeding, he needed to take more seriously than we do the problem it presented. He simply did not have our luxury of disregarding the limits of change by artificial selection.

Nevertheless, we might still worry how distinct these various functions were in the actual attitudes toward evolutionary theory. Wouldn't the heuristic uses of domestic breeding make belief in the theory more likely, merely by making it easier to understand? Actual attitudes and beliefs are messier and fuzzier than philosophical reconstructions typically suggest. This might well be true; we are plausibly more likely to believe theories that we understand. But there are still reasons for us to be clear about these different functions, even if, in practice, they are not always distinct. If we did not distinguish the heuristic from the evidential, we would miss an important problem for Darwin and how he responded to it. The analogy with domestic breeding was useful in various ways, but counterproductive in other ways. How is it possible to use it in a supporting manner without also allowing a counterproductive use? And if we do not distinguish these different functions, we simply miss the quandary that confronted Darwin. This is the flaw of the causal efficacy view. It does not recognize this quandary and its significance. By making these distinctions we are better able to discover the complexities in Darwin's epistemic situation. As useful as domestic breeding was for Darwin, its value was not unequivocal.

ACKNOWLEDGMENT

I am indebted to Peter Achinstein for his helpful criticism, guiding hand, and friendship. Although this chapter has benefited from both his philosophical insights and his critical attention, the views presented—and mistakes—are my own.

NOTES

1. I will refer to this as the "causal efficacy" interpretation, first because this is the term I used in two previous papers on the subject, and second because it distinguishes the use of the causal analogy between artificial and natural selection to establish the power of natural selection, from the use of the causal analogy to facilitate understanding. The term "causal analogy" applies to both functions.

233

2. Achinstein distinguishes *reasons to accept* and *reasons to believe* a hypothesis. Reasons to believe, according to Achinstein, are always epistemic reasons. Reasons to accept may include purely pragmatic reasons—the usefulness of accepting the hypothesis. What is important to scientists, he claims, is whether there is a good (epistemic) reason to believe a hypothesis. He therefore identifies belief as the important factor, not acceptance. See Achinstein, 2001: 24. I will follow Achinstein in focusing on *reasons to believe* as most relevant to questions about evidence.

3. Achinstein rejects three weak concepts of evidence. On the "Bayesian" concept e is evidence for h if and only if e raises the probability of h—even if only very slightly. On the "h-d" concept, e is evidence for h if e is deducible from h. And on the "satisfaction" model, "an observation report is confirming evidence for a hypothesis, if the hypothesis is satisfied by the class of individuals mentioned in that report." The report that a raven is black is evidence for the hypothesis that all ravens are black, and so is the report of a non-black non-raven, because the hypothesis is "satisfied" by both reports. See Achinstein, 2001: 6.

4. On this interpretation, the truth of h is a necessary condition for veridical evidence but not for potential evidence (Achinstein, 2001: 28).

5. Some of my analysis, in this and other sections, is found in more detail in Richards, 1997.

6. This formulation of the argument is ambiguous; it does not distinguish the different kinds of evolutionary change. I have left it ambiguous because that is typically how it is formulated. I will distinguish the various senses below.

7. See also Ruse, 1999: 178–79; Oldroyd, 1986; Hodge and Kohn, 1985; Hodge, 1993; and Waters, 1986. For a more recent account, see Sterrett, 2002.

8. The terms in this argument—"permanent change" and change able to produce new species—are not those of the causal efficacy argument as stated above: "great change." I will clarify these various ways of characterizing change, but what is important here is that establishing the efficacy to produce merely temporary change is not sufficient to establish the efficacy to produce the required kind of change.

9. Ruse argues that this is in fact the case (Ruse, 1999: 203–4). See also Bartley, 1992; and Beddall, 1998.

10. In Richards (1997), I argue that this emphasis on the differences rather than the similarities is good reason to reject the view that Darwin is making an analogical argument. The emphasis on the differences suggests that he is making *what he would regard* as experimental or inductive inferences.

11. Sterrett, 2002: 157 argues that for Darwin, artificial and natural selection are not always acting in opposition. She is correct in that one *can* select for traits that enhance fitness—speed in horses and dogs for instance, strong bones in cattle, and so forth. However, by protecting the favored individuals from competition and environmental demands, a breeder will be allowing harmful traits to persist and accumulate. So artificial and natural selection can sometimes operate "hand in hand" in a very narrow and limited manner, yet still ultimately be in opposition.

12. In a population of giraffes, for instance, even if long necks were favored by natural selection, change would still be limited if there were no cases of increasingly elongated necks appearing in each new generation. Neck length would not increase.

13. In this sense, facts about change under domestication can function to support a set of auxiliary hypotheses about heredity and variability that make the causal efficacy of natural selection plausible. Notice this indirect support is not analogical; it is support for laws that operate identically under domestication and nature.

14. Asa Gray, Chauncey Wright, and others understood Darwin to be making this experimental use of domestic breeding (Richards, 1997).

15. See Richards, 1997 for further analysis.

16. In his notebooks, Darwin distinguished two great systems of natural laws—inorganic and organic (Barrett, Gautrey, et al., 1987: 611).

17. This function of domestic breeding has been partially recognized by Ruse, Evans, Bartley, Rheinburger and McLaughlin, and others. But there is a general failure to specify Darwin's reasoning and the significance he places on these laws.

18. See Richards, 1997.

19. I am adopting this idea from Achinstein, 2001: 107.

20. For this point and those that follow there are many sources, but see Mayr, 2001 for one example.

21. Other disregarding evidence comes from work on guppies and finches (Zimmer, 2001: 86–87).

22. We might, for instance, find reasons to disregard facts about change under domestication in how we now conceive of species, and in our knowledge of the molecular basis of development and heredity. How these might work is worth investigating, but beyond the scope of this chapter.

23. That is not to say that we could not—if we wished—try to determine if the analogy with domestic breeding really should be regarded as counterproductive to the establishment of the causal efficacy of natural selection. If so, then strictly speaking, we would not be *disregarding* facts under domestication, but reinterpreting them. Modern evolutionists, however, do not seem to pursue this project.

24. That Darwin looked for evidence in nature, suggests that he did not regard the analogy as providing sufficient reason to accept the hypothesis. It does not suggest that he *could not* have regarded the analogy as providing evidence, however.

25. This analogy is therefore not what Achinstein calls an identical property analogy (Achinstein, 1991: 221).

26. Sterrett, 2002 argues this for this function, and I discuss her views further below.

27. He might, it is true, discuss the differences between the analogues in order to show that they are not real differences. For instance, he could argue that because artificial selection has had much less time to operate, we should not expect it to be as efficacious as natural selection—and that if it did have as much time it would have done as much. But Darwin is not arguing here that these are not real differences. He could not be more clear that these are real and important differences.

28. Methodological selection is selection by a breeder with the purpose of producing a certain kind of change. Unconscious selection is selection by the breeder without that purpose.

29. She does not, however, distinguish the different senses of causal efficacy—to produce change, unlimited divergence, new species or genera, and so forth—so it is unclear what sort of causal efficacy the analogy is supposed to establish.

REFERENCES

Achinstein, Peter, *The Book of Evidence*, Oxford University Press, NY, 2001.

Achinstein, Peter, *Concepts of Science*, The Johns Hopkins Press, Baltimore, 1968.

Achinstein, Peter, *Particles and Waves*, Oxford University Press, New York, 1991.

Barrett, Paul H., P. Gautrey, S. Herbert, D. Kohn, and S. Smith, eds., *Charles Darwin's Notebooks*, 1836–1844, Cornell University Press, Ithaca, NY, 1987.

Barrett, Paul H., and R. B. Freeman, eds., *The Works of Charles Darwin*, New York University Press, 1987.

Bartley, M., "Darwin and Domestication: Studies on Inheritance," *Journal of the Story of Biology*, 25, 1992.

Beddall, B. "Wallace's Annotated Copy of Darwin's Origin of Species," *Journal of the History of Biology*, 21, 1988.

Browne, Janet, *Charles Darwin: Voyaging*, Princeton University Press, Princeton, NJ, 1996.

Burkhardt, Frederick, ed., *Charles Darwin's Letters*, Cambridge University Press, 1996.

Burkhardt, Frederick, and S. Smith, eds., *The Correspondence of Charles Darwin*, Cambridge University Press, 1988.

Darwin, Charles, *On the Origin of Species*, facsimile of the First Edition, Harvard University Press, Cambridge, MA, 1964.

Darwin, Charles, *The Variation of Animals and Plants Under Domestication*, 2nd edition, New York University Press, 1988.

Darwin, Frances, ed., *The Autobiography of Charles Darwin and Selected Letters*, Dover, New York, 1958.

Darwin, Frances, ed., *More Letters of Charles Darwin*, D. Appleton and Co., New York, 1903.

De Beer, Gavin, *Evolution by Natural Selection*, Cambridge University Press, 1958.

Evans, L. T., "Darwin's Use of the Analogy between Artificial and Natural Selection," *Journal of the History of Biology*, 17, 1984.

Gray, Asa, *Darwinian*, Harvard University Press, Cambridge, MA, 1963.

Hodge, M. J. S., and David Kohn, "The Immediate Origins of Natural Selection," *The Darwinian Heritage*, Princeton University Press, Princeton, NJ, 1985.

Hodge, M. J. S., "Discussion: Darwin's Argument in the Origin," *Philosophy of Science*, 57, 1993.

Hull, David, *Darwin and His Critics*, Harvard University Press, Cambridge, MA, 1973.

Huxley, Thomas H., *Darwiniana*, D. Appleton and Co., New York, 1896.

Lyell, Charles, *Principles of Geology*, University of Chicago Press, 1990.

Mayr, Ernst, *What Evolution Is*, Basic Books, New York, 2001.

Oldroyd, David, "Charles Darwin's Theory of Evolution: A Review of our Present Understanding," *Biology and Philosophy*, 1, 1986.

Recker, Doren, "Causal Efficacy: The Structure of Darwin's Argument Strategy in the Origin of Species," *Philosophy of Science*, 54, 1987.

Rheinberger, H. J., and P. McLaughlin, "Darwin's Experimental Natural History," *Journal of the History of Biology*, 17, 1984.

Richards, Richard A., "Darwin and the Inefficacy of Artificial Selection," *Stud. Hist. Phil. Sci.*, 27, 1997.

Richards, Richard A., "Darwin, Domestic Breeding and Artificial Selection," *Endeavour*, 22 (3), Elsevier Science Limited, 1998.

Ruse, Michael, *The Darwinian Revolution*, 2nd edition, University of Chicago Press, 1999.

Sterrett, Susan G., "Darwin's Analogy Between Artificial and Natural Selection: How Does It Go?", *Studies Hist. Phil. Biol. & Biomed. Sci.*, 33, 2002: 151–68.

Wallace, Alfred R., *Darwinism*, MacMillan & Co., London, 1989.

Waters, Kenneth, "Taking Analogical Argument Seriously," *Proceedings of the 1986 Biennial Meeting of the Philosophy of Science Association*, Vol. 1.

White, Michael, and John Gribben, *A Life in Science*, Dutton, Penguin Books, New York, 1995.

Zimmer, Carl, *Evolution, Triumph of an Idea*, Harper Collins Publishers, New York, 2001.

 12

Evidence in the Sciences of Behavior

Helen E. Longino

The behavioral sciences—psychology, sociology, ethology—have been characterized by multiple schools, multiple approaches, multiple theories. So much is this so, that some are tempted to dismiss these sciences as either not science at all or as at best pre-scientific, lacking as they do a single paradigm. But the multiplicity of approaches and consequent debates about them cannot so easily be avoided, because the debates concern what is to be measured or modeled, and how and why. The label *pre-scientific*, furthermore, is misleading. There is a great deal of activity qualifying as science by any definition that is occurring throughout the broad range of behavioral sciences: various kinds of observation, measurement, experiment, and hypothesis evaluation. The source of discord may lie partly in the complexity of the phenomenon, rather than in the immaturity of the research. It may also lie in false expectations: the hope that there is a decidable empirical basis for all policy questions and the hope that evidence can be decisive regarding all issues in the scientific study of behavior.[1]

Research on both causal and compositional questions can proceed in a top-down direction or a bottom-up direction. Top-down research identifies a behavior of interest (e.g., aggression); characterizes the behavior in a way that makes it amenable to study; finds appropriate measures of the behavior; and then seeks correlates in lower-level behaviors or temperamental characteristics, and ultimately in biological factors such as neural states, non-neural physiological states, or genes. Bottom-up research studies the etiology of lower level factors, such as temperamental components (e.g., impulsivity) or neural states, neuro-active substances, and non-neural physiological states, and seeks correlates of these at the behavioral level. The challenge to researchers is to design measurement and observational protocols sensitive enough to pick out significant relationships and to differentiate between possible causal factors.

In the public eye, the big question about behavior is the nature versus nurture question: Is behavior B a result of our genes or our upbringing? Although it looks as though there is one question that multiple approaches are asking—what is the cause of x?—on closer examination there are quite different kinds of causal question that are asked within the different approaches. The debates among proponents of different approaches, even when articulated as debates about different causal theories, are at least as much, if not more, about the point or value of different kinds of knowledge.

In this chapter, a prolegomenon to a more extensive analysis of this research, I wish to dispel the idea that there is an empirical way of settling disputes among proponents of the different approaches, by showing that each approach generates its own set of questions, to which different forms of data can provide answers. Even though some hypotheses supported within the different approaches seem to be in contradiction, the conflict is not, or not always, one that can be settled by appeals to evidence. Instead, the disagreements are best understood as conflicts between approaches. At the minimum, an approach includes characteristic questions, characteristic methods for addressing those questions, and a commitment to the importance of the questions and the answers generated by the available methods.

First, I describe each of a set of approaches to studying behavior in a general way, noting whatever explicit theoretical framework is adopted and what kind of knowledge is sought. I then look more closely at the questions and methods of answering them to consider the import of the available and possible evidence. Finally, I suggest an alternative way to cast the disagreements between the advocates of the different approaches.

FOUR APPROACHES TO STUDYING BEHAVIOR

Behavior Genetics

Behavior genetics includes both quantitative (or classical) behavior genetics (the application to behavior of population genetics) and molecular behavior genetics (the application of molecular genetics to behavior). Granting that both the genes and the environment of an organism causally influence its behavior, the behavior geneticist asks what the genetic contribution to a given behavior is. The main tool that is used is analysis of variance, that is, identification of how much of the difference in expression of a trait in a population is correlated with genetic difference. This requires being able to separate or distinguish differences in environment from differences in genotype, as well as to specify what counts as sameness.

In humans, the methods traditionally used have been twin studies and adoption studies, in which one or another of the contributing factors is purportedly held constant in order to measure differences in the other. Researchers have compared the degree of similarity (concordance) in expression of a given trait by monozygotic (MZ) twins who share an identical genotype and by dizygotic (DZ) twins who share only half of their genes. The assumption is that the environment is the same for the members of the DZ twin pairs, and that only the genotype differs. A finding of higher concordance in a trait among MZ than among DZ twins is treated as data supporting claims of heritability of that trait. Researchers have also compared twins reared apart with twins reared together. Here the assumption is that the environments differ for the separated twins and not for the twins reared together, whereas the genes in both kinds of pair remain the same. In adoption studies, researchers compare adoptees with both their biological and their adoptive parents, and sometimes with biological and adoptive siblings. While most proponents and commentators understand quantitative behavior genetics as applicable to differences within populations, some have argued that it can also be used to determine the genetic contribution to individual development as well.[2]

A typical quantitative behavior genetics study of aggression measures concordance in some indicator of aggressive behavior, such as felony conviction or antisocial personality diagnosis, among biologically related family members in comparison with non- or less-biologically related family members, whether MZ versus DZ twins or biological versus adoptive parents and siblings.[3] Researchers Mason and Frick identified 70 such studies of aggressive behavior (that did not involve confounding issues such as alcohol abuse) published between 1975 and 1991.[4] They developed a set of selection criteria that the studies had to satisfy in order to be included in a meta-analysis. They then sorted the behaviors reported in the studies into three categories: criminal offending, aggression, and (other) antisocial behavior. Each of these was then divided into the categories "severe" and "non-severe." According to Mason and Frick, the studies supported the conclusion that there are genetic effects on antisocial behavior, and that these are stronger for severe antisocial behavior than for the non-severe.

Behavior genetics has recently acquired another research tool, thanks to molecular genetics: linkage analysis. Linkage analysis looks for common genetic markers in biological relatives identified as sharing the same behavior. The markers are known multi-allelic loci on a given chromosome. The strategy is to associate allelic variation with behavioral variation. If such an allelic association can be found, that is taken as evidence that a gene in the same region of the chromosome is involved with the behavior.

In 1993, 14 male volunteers of a Dutch family, all of whom experienced episodes of aggressive behavior, were found also to share genes on the X chromosome coding for a particular enzyme: monoamine oxidase, or MAOA.[5] Also in 1993, Dean Hamer and colleagues at the National Institutes of Health announced that pedigree analysis showed an elevated incidence of homosexuality in brothers of gay men (13.5%) and in maternal uncles and sons of maternal aunts of gay men (7.5%).[6] This incidence, given the size of the sample, is significant if one assumes a background rate in the general population of 2%. The pedigree analysis suggested X chromosome involvement. Linkage analysis revealed common markers at the Q28 locus of the X chromosomes in 33 of 40 sibling pairs selected from the larger sample. As with twin and adoption studies, the interpretation of results like these is controversial. Behavior geneticists interpret the twin and adoption studies and linkage studies as indicating that there is a significant heritable (and when backed by linkage analysis, genetic) component in the behaviors studied.

Human behavior genetics is buttressed by a variety of studies on non-human animals. In a research process known as reverse genetics, a form of bottom-up research, researchers introduce mutations into the genome and identify the consequent variations in behavioral routines. These studies, performed largely on organisms such as varieties of *drosophila*, can be used to suggest mechanisms of gene expression, as well as behaviors in other organisms (e.g., humans) possibly subject to genetic influence.

Proponents of behavior genetics claim on its behalf that it can help elucidate the mechanism's underlying behavior; that it represents the appropriate extension of evolutionary—Darwinian—concepts to behavior; that it can help show the role of non-genetic, environmental factors in behavior; and that it can indicate the limits of various kinds of intervention strategies, as well as identify the populations that may or may not benefit from such strategies.[7]

Social-Environmental Approaches

The social-environmental approach emphasizes the environmental contribution to the development and expression of various behaviors.[8] Socially or environmentally oriented studies seek to establish the role of socialization patterns, familial and school environments, peer relations, media exposure, and/or parental attitudes and interactions with their children. The methods used include correlational studies (using information gleaned from various public records and from interviews and surveys) and the direct observation of behavior and interactions in standardized (laboratory) settings. Typically, the subjects in these studies are young children or adolescents, although in some cases the adult behavior of persons identified as having had a certain kind of experience in childhood is also studied. Path analysis is used when many variables are being examined. In some cases, researchers conduct longitudinal studies, following individuals for six or more years. The young and adolescent subjects for these studies are identified in schools or through clinics.

In one such study, researchers sought to correlate familial interaction patterns with long-term disruptive behavior in eight- and nine-year-old boys. The boys were identified by teachers, who were asked to complete Social Behavior Questionnaires on their students. Interactions in 44 families were studied by observing the parents and child in question engaged in joint tasks in the researchers' laboratory. Observers used checklists in rating dyadic interactions between father and child, mother and child, and between the parents. Researchers found that negative behaviors (such as verbal abuse or attacks) and positive behaviors (such as endearments) in the parent-child dyads were not reciprocal, but that negative behavior of one parent toward the boy was correlated with negative behavior on his part toward the other parent. In addition, negative behavior of boys toward their mothers was correlated with fathers' negative attitudes toward their female spouses.[9]

Until recently, a great deal of the literature in psychology and sociology on social-environmental determinants of homosexual orientation was articulated in the context of a disorder or deviance model of homosexuality. (Homosexuality was not removed from the American Psychiatric Association's list of psychiatric disorders until the 1970s.) Accordingly, this literature has explored such factors as the overprotective mothering of boys and has focused on sexual behaviors in Western societies. Anthropological cross-cultural studies offer another perspective, as well as an alternative to the individualistic approaches to behavior found in other approaches. Barry Adam, reviewing a number of studies, proposes that in certain social configurations, homosexuality is an outgrowth of particular combinations of age, gender, and kinship structures.[10]

Practitioners of social-environmental approaches have various views about what can be concluded from such research and what the point of it is. Some researchers clearly assume a causal effect of parental behavior on children.[11] Others acknowledge that in the networks of association their studies uncover, causal relations might go differently than they assume them to.[12] Diana Baumrind takes something of an equivocal (perhaps strategic) position: causal inferences are not licensed by the correlational data produced in these studies.[13] Such data are valuable for eliminating, but not confirming, hypotheses. Nevertheless, she claims, it is reasonable to conclude on the basis of a number of studies that parental behavior does influence children's development. This being the case, these environmentally oriented studies are valuable because they can (help to) identify children "at risk" of antisocial behavior or delinquency and identify points of intervention in family and school dynamics. And in a clearly melioristic declaration of faith, she says that even if heritability accounts in large part for what is, it doesn't follow that social interventions can't improve on that. The implication is that social-environmental research approaches are required to identify which social interactions are effective.

Developmental Systems Theory

The developmental systems approach has its theoretical base in embryology and developmental biology. Here the central question is how the organism develops from a single fertilized cell into a mature individual characterized by multiple and specialized organs and tissues. Differentiation, the process of specialization of cells, is one of the key problems for developmental biologists. For the systems approach, genetic and environmental contributions to development are not separable. The relation between them is non-additive and nonlinear. Nor can behavioral development be separated from the other dimensions of development.

Gilbert Gottlieb, one of the principal researchers in this tradition, characterizes development as emergent, coactional, and hierarchical.[14] By hierarchy, Gottlieb seems to mean the multilevel character of development and the interlevel, as well as intralevel character of coaction. "Hierarchy" is not used to convey dominance or causal priority, but rather the unity of genetic, physiological, neurological, behavioral, and environmental aspects of the developmental system. It is being used, therefore, much as Grene and Eldredge (1992) used it to represent something like degrees of embeddedness, or of enclosure.[15] By "emergent," Gottlieb means that structural and functional complexity increases at all levels of organization of the individual as a result of interactions within and between these levels. By "coactional," Gottlieb means that the factors involved in development interact. Not only do these factors not act independently, but they can also modify one another, thus altering their respective contributions in subsequent phases of development.

A variant of the Developmental Systems approach is the Dynamic Systems approach. The difference, if any, lies more in emphasis. Whereas the Developmental Systems advocates stress the complexity of interaction of endogenous and exogenous factors, the Dynamic Systems advocates stress their temporal nature: the dependence of one stage of development on earlier stages. Esther Thelen's research offers one of the few examples of the application of this approach to the study of human behavior.[16] She has studied the development of basic motor skills, such as reaching and grasping in infants, skills that involve cognitive, neural, and muscular coordination. Her research involves systematically varying elements of the situation implicated by the components of such skills by, for example, changing the contents of the perceptual environment, altering the speed at which objects move in that environment, measuring arm angles and wrist torque, altering constraints on muscles (with weights), and repeating these variations over time to see how the organism learns to integrate component capacities into coherent behaviors.

Developmental systems theorists see the point of their work as producing a comprehensive understanding of the mechanisms of development. The object of their interest is the individual organism, not populations. For them, the core questions are those of differentiation. In the realm of behavior, this means understanding how some behaviors are canalized—that is, fixed—while others remain malleable throughout an individual's lifetime. Researchers seem less concerned about practical applications such as interventions, although some have argued that there are implications of taking the systems view. Lerner argued that a systems or contextual approach, because it involves a finer attention to differences between individuals and between environments, requires abandoning notions of the "generic child" (which tends to be a white middle-class child) in the design of social intervention strategies.[17] Instead, researchers and clinicians should take into account contextual variability (both variation in developmental contexts and variation in the interactions of different individuals in the same context). This insistence on variability and context dependence means that, if generalizations are to be obtained within this approach, they will not be generalizations about populations, individuals, or their properties, but generalizations about processes. Such generalizations will be about more abstractly conceived entities such as canalized behavior, rather than about aggression or sexual orientation per se.

Neurophysiological and Neuroanatomical Approaches

Neurophysiological and neuroanatomical approaches to behavior seek to identify and characterize aspects of the neural substrate of behavior. The Decade of the Brain (1990–2000) in the United States encouraged and funded a great deal of research on the human brain and nervous system. Thus much more is known about the brain than was known before. Nevertheless, it's not clear that our overall understanding of brain function has kept pace, since that would require integrating the disparate bits obtained by the methodologies such as electroencephalography and the newer imaging technologies. Although there are some general theories of brain function, such as the neuronal group selection theory advanced by Gerald Edelman, that drive experimental programs, the latter tend to be either research on cognitive function or research at the molecular level.[18] Research on neural correlates of aggression and of sexual orientation does not seem part of any specific theory of brain function, but is carried out under the not-unreasonable assumption that accounts of brain and neural function will be part of any complete account of the mechanisms of behavior.

There is a vast literature detailing experiments with laboratory animals with respect to both aggressive and sexual behavior. Recent work on humans takes advantage of new imaging technologies such as PET (positron emission tomography) and NMRI (nuclear magnetic resonance imaging), as well as new developments in neurochemistry and neuropharmacology. In addition, some studies use traditional methods, such as postmortem dissections, for structural information. Work on laboratory animals involves manipulating levels of neuroactive substances and observing the effects of those manipulations on animal behavior.

As an example of animal studies, the laboratory of Craig Ferris has been studying the role of vasopressin and vasopressin antagonists in enhancing or decreasing the expression of highly stereotyped forms of aggressive behavior in several rodent species.[19] Here the interest is in understanding the functions of vasopressin in the organism. Not only behavior, but also internal effects such as binding to specific sites in the hypothalamus are studied. Stereotyped behaviors, such as flank marking, are good indices of behavioral effects, just because their stereotypical nature makes them easy to observe and identify. Thus aggression is of secondary interest in this research—any behavioral effect that is easy to study would do as well.

When the research involves human subjects, the situation is not so straightforward. Serotonin, a neurotransmitter, was first studied in non-human animals, rodents, and later rhesus monkeys.[20] There it was found to be inversely related to irritability—understood as a tendency to respond aggressively to adverse stimuli—which suggested a similar role in humans. The role of the serotonergic system—the process of neurotransmitter release and reabsorption (metabolism)—in a variety of dysphoric conditions has been studied via three primary methods in humans. Concentrations of serotonin metabolites can be measured in cerebrospinal fluid, with reduced concentrations of the metabolites signaling decreased serotonin activity. Also, there are strategies for indirectly measuring the presynaptic sites of serotonin reuptake (*reuptake* means lower levels of serotonin activity). And finally, "pharmacochallenge" studies involve administering pharmaceutical agents, especially serotonin agonists or mimics, to study serotonin activity in the brain (e.g., the specificity of receptor sites).[21] All three of these methods support the conclusion that diminished serotonin activity is associated with (impulsive) aggressive and antisocial behavior.

Neuroimaging strategies are used to identify areas of greater or lesser activity in the brain. Computerized tomography and magnetic resonance imagery provide structural information. The former uses X-rays passed through an individual's head from multiple positions to create an image. The latter disturbs hydrogen ions with radio waves and employs a magnet to realign the ions. Differences in resonance frequencies emanating from different tissues in the brain are then used to create the image. Regional cerebral blood flow (rCBF) studies use inhaled or injected xenon or similar tracers to detect areas of enhanced blood flow and, by inference, increased activity. Positron emission tomography (PET) assesses rates of glucose metabolism by means of a radioactive tracer that leaves a residue, which can be detected by a scanner, at sites of metabolic activity. In both MRI and PET the better resolution is obtained from the much more costly instrument, thus limiting the sample sizes in those studies. And the data obtainable from the different strategies are not always correlated, so any information gained by these strategies must be regarded as preliminary.[22]

Nevertheless, researchers are attempting to link dysfunction in particular brain areas with particular forms of aggressive or assaultive behavior. For example, Adrian Raine and colleagues used PET scanning to study the brains of persons charged with murder who were referred to psychiatric clinics in preparation for insanity defenses or a determination of competency to stand trial.[23] They found diminished prefrontal cortical function in that group, as compared with matched controls, and similar levels of function in other brain areas in the two groups, leading them to conclude that prefrontal cortical dysfunction could be implicated in violence in some offenders. They caution that the sample was small and consisted of individuals already suspected of mental illness or organic brain injury.

Both functional and structural studies have been performed in connection with sexual orientation. In the 1990s, Simon LeVay reported on a comparative study of brain structure in heterosexual and homosexual patients who had died of AIDS.[24] LeVay chose to focus on the medial preoptic area of the anterior hypothalamus, because that area had been identified as involved in sexual behavior in rats. Although he found that one set (of four) of the nuclei in the homosexual men's brains were closer in size to those of the (undifferentiated with respect to sexual orientation) women's brains than to those of the heterosexual men's brains, those nuclei are not ones identified as sexually relevant in the rat studies. The significance of these findings is thus not known.

Quite obviously, the point of studies of these kinds is better to understand the neural substrate of human behavior and action. In this regard they can be seen as part of the much larger research program of the neurosciences, but focused on understanding the function of particular aspects of the brain and nervous system, such as particular neurotransmitters or particular structures. The large gap between measurable changes in these features and changes in particular associated behaviors, however, means that this work is in its very early stages. Furthermore, although in some cases (e.g., injury that alters structure or function, followed by a change in behavior) the direction of causality can be fairly confidently affirmed, in others the causal relations among empirically associated phenomena are still unknown. This is true of most of the work on aggression and of the work on sexual orientation. Since in both cases the functional and structural differences could be consequences of behavior that has been canalized by other factors, its causal role cannot be unproblematically inferred.

246

But the work done specifically on aggression also has a variety of pragmatic purposes. The work on serotonin, for example, informs psychopharmacological interventions such as the administration of Prozac or other agents to aggressive individuals. It thus is implicated in both medical and commercial networks, with corresponding social and political reverberations.[25] This work, as well as the neuroimaging and physiological research, also suggests the need to distinguish among subcategories of aggressive individuals, some of whom may have biologically based dispositions that are either toward aggression or will be expressed as aggression under certain conditions. Low-impulse control, for example, crops up in a number of studies as a differentiating feature associated with various functional or structural differences.

PRELIMINARY ANALYSIS OF THE STRUCTURE OF INVESTIGATION

Comparative examination of the studies carried out in these different families of investigation shows that distinctive questions stimulate the development of distinctive methodologies. Here is a more systematic presentation of these components of the respective investigations.

Questions

What appear to be the common or umbrella questions—What causes behavior? What is the most significant factor in the fixation of behavioral dispositions?—cannot be directly answered. Questions about behavior must, therefore, be more carefully and precisely articulated. They must also be articulated in the vocabulary of the state of the art at the time. The consensus is that both genes and environment, or both endogenous and exogenous factors, influence behavior. This means that a number of questions can be generated, each of which in turn produces a cascade of further questions. Each of the approaches discussed can be distinguished by a single question from which others follow, as in the list below:

What role(s) do genes play in behavior B?

- How much of B is genetically influenced?

- To what degree is B heritable? (How much are differences in parents correlated with differences in offspring? How much do differences in parents influence differences in offspring?)

- How much of the difference in expression of B in a population is associated with genetic difference?

- Does the degree of genetic influence on B change over time?

- How can the methodologies used to study the genetic influence on behavior (twin and adoption studies) be refined and extended?

- Can any genetic markers be associated with the incidence of B in a given pedigree?

- Can linkage analysis be refined and extended to strengthen claims of genetic influence?

What role do environmental and other exogenous factors play in behavior B?

- What role do gross- or macro-level social variables (such as social class; ethnic, racial, and cultural identity; urban/suburban/rural orientation; immigrant/native status) play in the expression/frequency of B? Does one or more of these predominate in the expression of B?

- What role do micro-level variables such as family, school, peers, or media exposure play in the expression of B? Does one or more of these predominate in the expression of B?

- Does the influence of micro-level variables in the expression of B vary in relation to macro-level variables? How?

- How do differences within a family influence the expression of B by its members?

- How can familial interactions relevant to B be studied?

How does B come to be expressed in individuals?

- What developmental trajectories can be identified that culminate in B?

- What developmental factors (genetic, epistatic, intrauterine, physiological, physical and social environment) interact in the development of B?

- Is the disposition to B canalized? If so, how?

- At what levels of organismic integration and organization do the causal/developmental processes relevant to B occur?

- Does the development of species-typical traits differ from the development of individually variable traits? (Of which type is B?)

- How do complexity of organization and specialization of function develop?

- How can intralevel and interlevel interactions be studied?

What role do neural structures and processes play in behavior B?

- Can a specific, local neural structure or process be associated with occurrences of B?

- Are the neural processes associated with B distributed or local?

- How are the processes associated with B activated? How are they inhibited?

- How do the processes associated with B interact with other neural and organic processes?

- How can available techniques be improved to study the role in the expression of B of the neural processes associated with it?

Each of the listed subquestions also generates additional questions as researchers go about answering them. Although this is especially true of the methodological questions, the substantive questions are also refined as research proceeds. There are, of course, questions that cross approaches. One could ask, for example, how genetic and environmental factors interact, or how neural and other factors interact. Only the developmental systems theorists attempt explicitly to address questions about interaction, and then only in the context of developmental questions. The questions just listed, however, can be pursued independently of whatever progress or lack of it is occurring in the other approaches. Each approach becomes self-sustaining through the modification of its questions in practice and by the consequent generation of more and more refined questions. The research programs derived from an initial question are thus driven further into specificity and autonomy.

249

Methods

One of the critical methodological issues is the identification, individuation, and definition of behaviors and of the causal factors contributing to their expression. A second issue is how to measure and establish the relatedness of the phenomena under investigation. Here I focus on the second of those issues.[26]

Behavior Genetics

Although there is a common overall question (thesis) as noted above, behavior genetics now encompasses at least two sorts of approach, each with a distinctive set of methods. Classical human-behavior geneticists use traditional methods of ascertaining heritability: twin studies and adoption studies. More powerful experimental methods, such as the breeding used in classical fruitfly genetics, are not appropriate for human studies, for reasons of both ethics and scale. Although they are carried out under the rubric of genetics, what twin and adoption studies can show at best is the extent of heritability of a trait in a given population, in a given common environment.

So the question to which twin studies can provide answers is, "How much of the difference in the expression of B in environment E is heritable?" The assumption is that "heritable" here is equivalent to "genetically heritable." Alternative or competing hypotheses would assign different heritability quotients. The studies cannot, however, be generalized to answer the question, "How much of the difference in the expression of B is heritable?" Both the variability and heritability of a trait can change when environments change. Thus, critics emphasize that twin studies do not enable conclusions about causation in individual organisms, but only about distributions of a behavioral trait in a population in a given environment. Multivariate analysis can assist in the decomposition of complex traits, and longitudinal studies can help in identifying shifts in the relative influence of heritable and non-heritable factors over time. But twin studies are not capable of distinguishing between intrauterine and genetic effects. And because all that is known about twins is that they share a genome, they are not capable of distinguishing between polygenic and monogenic traits, nor, obviously, are they able to identify genes.

This is where molecular genetics steps in. So-called *reverse* genetics introduces specific mutations at specific sites of the genome. Applied to the behavior of fruitflies, this research strategy makes possible the identification of allelic variants that are strongly associated with behavioral variations. Although such manipulation of human genes is not possible, retrospective identification, through linkage analysis, of loci associated with a phenotypic trait is possible. Here the strategy is to find genetic markers—alleles with a high degree of variability. If variant alleles (markers) match variants in a behavior, the inference is that a gene in the vicinity of (linked to) the marker is involved in the behavior in question. Molecular studies can, in principle, identify genes or alleles whose presence in the genome plays a major causal role in the expression of a phenotypic trait.[27] The markers, of course, are not the genes themselves, and more refined work is required to identify the particular sequence rather than the region of the chromosome where the sequence is located.

One hypothesis was suggested by the earlier mentioned study of a Dutch family, a large percentage of whose male members had high levels of aggressive behavior. The common pedigree prompted a genetic investigation, which showed that they shared a genetically based deficiency in monoamine oxidase A. Alternative hypotheses would concern the possibility that the common behavior was independent of the shared MAOA deficiency. The finding suggested that MAOA deficiency might be implicated in some cases of aggression, but since the deficiency is not nearly as widely distributed as aggressive behavior, no general hypotheses about the etiology of aggressive behavior are suggested.

Social-Environmental Approaches

The social-environmental approaches use measures of aggression similar to those used by the behavior geneticists, except that some studies treat a continuum of social behaviors from the antisocial to the pro-social, rather than limiting themselves to the antisocial. The pro-social (or "sociability") behaviors include offers of assistance, participating in cordial verbal exchange, and display or exchange of affectionate gestures. Aggression is measured by physical (hitting and starting fights) and verbal behavior (angry, hostile speech) as

250

ascertained by self- and other-reports; delinquency as ascertained in court records; psychological classifications such as antisocial personality disorder and childhood conduct disorder, as ascertained through psychiatric diagnosis; or hostile, confrontational interactions ascertained through direct observation.

The social-environmental approaches in psychology seek to associate distributions of one or another of these measures of aggression with variation in some environmental factor, such as parental behavior, educational experience, or peer relations. Most of the studies using questionnaires or interviews report using more than one measurement strategy (both self-report and other-report, or both self-report and court records) to enhance the reliability of behavioral ascriptions. The methods employed are both retrospective and prospective. Retrospective methods are employed with populations whose relevant parameters are determined via interview and questionnaire, or via direct observation. Prospective methods involve introducing some change into some portion of the target population and determining its effect. Whereas controls in the retrospective studies are populations of individuals who do not exhibit the behavior under study, controls in prospective studies are individuals who exhibit the behavior under study, but who have not experienced the intervention in question.

What these studies can show is the relative strength of a particular social or environmental factor in comparison with others in the same general setting. They tend to assume either uniformity or random variation of genetic or other individually internal factors (such as hormonal or neurotransmitter secretion patterns) and thus measure the relative strength of the various kinds of social factor—parental interaction, parental discipline, parental attitude and behavior toward other family members, peer support, school environment—in producing a particular behavior. Typical hypotheses studied, then, would be (a) from retrospective studies ("Parents who were themselves abused as children are more likely to engage in emotional or physical abuse of their children") or (b) from prospective studies ("Coaching parents to adopt communicative style C in situations S reduces the frequency of aggressive responses from children"). Contrastive hypotheses would concern alternative potential environmental factors. Since the influence of any such factor may vary, given variation in other environmental factors, and since the magnitude of their population level effects may vary as and if genomic distributions vary from population to population, their conclusions are as limited in generalizability as the conclusions of classical and molecular genetics.

Developmental Systems

Developmental systems researchers have concentrated their empirical work on animal models and on human infants and children. Consequently, they have less to produce by way of empirical results about behavioral dispositions manifest in adulthood or late adolescence. Nevertheless, individual researchers claim that this approach is the appropriate one to adopt, either for human behavior generally or for some specific behavior.[28] There is, then, no issue yet about defining or measuring behaviors. Given the holism that characterizes this approach, it is not clear how detachable or separable from the functioning of the entire human organism specific behaviors (such as "starts fights without provocation," "hits without provocation," etc.) would be, or what strategies researchers would use to individuate and measure behaviors. In animals, by contrast, various behaviors amenable to experimental intervention have been studied. The main form of argumentation, however, seems to be either conceptual or addressed to the shortcomings of other approaches and re-analyses of their data.

The experimental methods involve intervening in a developmental system to show that a given factor is (contrary to what might be expected from the perspective of another approach) essential to normal development of a trait or behavior. The exception here is Esther Thelen and her colleagues, who, in their research on human infants, systematically vary all features that are involved in a particular kind of performance, if they can ethically be altered. These fine-grained studies of very basic elements of behavioral repertoire both support the general thesis of complex interaction and suggest how much more difficult it will be to study higher-level behaviors by taking the same approach. A hypothesis regarding such higher-level behaviors would emphasize the involvement and mutual modification of genetic, physiological, and environmental factors in the fixation of particular behavioral dispositions, and alternatives might identify different particular (sets of) elements from these categories or the roles of different developmental stages.

One point emphasized by these writers is the distinctness of their concept of a reaction-norm from the behavior genetic concept of a reaction range. The reaction range idea is implied in certain behavior genetic articulations of the genotype–environment interaction. Some behavior geneticists have described this as the genotype setting upper and lower bounds on the expression of a trait across a range of environments. The reaction-norm idea, instead, is that each genotype is "associated with a characteristic pattern of phenotypic changes in response to alterations in the environment . . . [but] the rank order of individuals [regarding degree of expression of a trait] can change appreciably and uncontrollably under novel conditions."[29] The point here is that while across the variation in an environmental component in one setting a given genotype will characteristically be distributed into one order or ranking from lowest to highest expression, in another environment this order will not necessarily be preserved.[30]

Conversely, although one might be tempted to rank environments along degrees of nutritiveness or nurturance, for example, the genotypes will not respond as though to a continuum of reinforcement, but may instead be differentially activated by the different environments. Instead of *all* allelic variants increasing (or decreasing) in the intensity of expression of a trait when exposed to varied environments, some may and some may not. The implications are equally dire for behavior geneticists as for social-environmentalists, since neither genes nor environment can be thought of as acting independently. There can be no question of partitioning out the separate effects of genetic difference or of environmental differences on the expression of an individual phenotype. Thus the inference from heritability of a given measure of aggression in a population in a given environment (or set of environments) to genetic causality is as problematic as inferring from results in one setting that improving parenting skills will, in general, diminish the likelihood of delinquency in offspring.

253

Neurobiology

Many of the neurobiological studies on aggression have worked with human populations already clinically identified as antisocial. However, when studying the behavioral effects of psychoactive agents, they also use categories of physical or verbal aggressiveness similar to those employed by the other approaches. In the case of sexual orientation, self-report or Kinsey measures based on questionnaire or interview responses are used. The study methods employed are retrospective, concurrent, and prospective. Retrospective methods include the use of autopsies to identify neurostructural correlates of behavioral patterns attributed to the individual, and correlational studies of prison and hospital records to identify associations between brain injuries or other trauma (e.g., birth complications) and later criminal behavior. Here a typical hypothesis would be, "Damage to region R of subcortical area S increases the frequency of aggressive episodes in patients characterized by conditions C." The contrast would be between absence of damage to R or damage to other regions.

Concurrent methods include brain imaging to identify areas of brain activity related to certain thoughts or sensory stimuli, as well as measuring changes in other physical parameters (heart rate) on exposure to certain cognitive or sensory stimuli. Here a typical hypothesis might be, "Aggressive behavior is causally influenced by increased activity in region R of the neocortex." Contrasting hypotheses would concern lack of activity in region R, or activity in other regions of the neocortex, or the absence of effects of activity in R.

Prospective methods include animal experimentation to identify the effects on behavior of organizational or activational exposure to bio- and psychoactive substances, and clinical trials in humans to ascertain the physiological, psychological, and behavioral effects of neuroactive substances. Here a typical hypothesis would be, "Administration of fluoxetine to clients previously diagnosed with antisocial personality disorder does not increase the frequency of physically aggressive episodes." Contrasting hypotheses would concern administration of fluoxetine to individuals not so diagnosed, or the behavior of individuals classified as ASP who do not receive fluoxetine, or comparisons of the effects of fluoxetine with those of some other neuropharmaceutical.

Although the retrospective or the concurrent methods can answer questions about the strength of the correlation of a given neural structure or process or type of damage with a given behavior, none can as yet produce data that definitively establishes proof that one kind of neural structure or process causally affects some behavior. While relevant to establishing causal claims, the correlations alone neither establish the direction of causality nor rule out a common cause. The prospective methods do establish that a given factor plays a causal role in diminishing, enhancing, or otherwise affecting a given behavior, but the knowledge thereby gained is still quite crude. The mechanism of action needs to be identified by other methods, and the effects of psychopharmaceuticals are generally quite a bit broader than the intended one.

Discussion

Generating evidentially relevant data in any of these approaches requires, as in all sciences, skill and ingenuity. As each approach gains in sophistication, its experimental and observational methodologies enable it to address more fine-grained questions about the relative impact of the types of factor it studies. In spite of the attention still given to the nature–nurture question,[31] these increasingly sophisticated methods are not designed to discriminate between the approaches. The question, "Do genes or social factors play the greater role in human aggression?" is just too crude to be addressed by the available methods of observation and calculation. Proponents of the different approaches, of course, agree in principle, because they acknowledge that both are involved. Still, they hold out the hope that there is an unequivocal answer to the question about the relative contributions of different factors. To date, however, the structures of investigation preclude acquiring evidence that is relevant to this question. The kinds of data generated by the approaches provide evidence relevant to discriminating between hypotheses internal to the approaches, but not to discriminating between hypotheses from different approaches.

In spite of the fact that they do not have the resources to show that one among them is correct in comparison to the others, the critical interactions that take place among their proponents do have consequences for the understanding of behavior. One consequence is that investigative resources proper to each approach are sharpened as a response to challenge and criticism. A second is that the limitations of each approach are made evident by the articulation of questions that they are not designed to answer. These critical interactions, therefore, enable the refinement of methodologies, the clarification of concepts, and the design of experiments and studies to control for causal factors demonstrated by others. All this makes for more knowledge, which, judged by means of the evaluative tools available within each perspective, is also *better* knowledge.

Whether the knowledge is good knowledge or knowledge worth having in any more general context depends on assessment at a different level. One approach will probably help elucidate some of the genetic and biological mechanisms underlying behavior, may show the limits of environmental influence and of social interventions in certain settings, or may lead to new genetic technologies. Another will help sort out which kinds of social-environmental factors, if any, predominate over others in the development of particular behaviors, and which social-environmental interventions are effective in increasing or decreasing the frequencies of behavioral outcomes.

255

Another approach will contribute to the articulation of the complexity of developmental interactions and will possibly contribute to the articulation of guidelines regarding intervention strategies. Another will contribute to understanding the role of specific neural structures and processes in particular behaviors, as well as to the identification of pharmaceutical or surgical interventions that will increase or decrease, enhance or diminish, certain behaviors. Which goal will be emphasized at any given time will depend on (1) the relevance and connections that can be made to other research programs; (2) the relevance that can be established to social interests; and (3) the priority given to certain cognitive and social aims over others.

Contrary to most of those who argue on behalf of one or another of the approaches, these matters are not just questions of which approach is empirically correct, or even more likely to produce results. They concern the *kind* of knowledge we seek to acquire, and they must be settled by appeal to different standards than are involved in evaluations of the adequacy of the content of models and hypotheses to their intended object. Advocates of the different approaches must make a case for the value of the kind of knowledge their approach is capable of providing. But this involves quite different skills than those involved in the design of experiments and the evidential testing of hypotheses.

NOTES

1. Because research on the etiology of behavior is always research about some particular (category of) behavior, it's necessary when engaging in comparative analysis to pick one or two behaviors in order to track similarities and differences across a consistent base. For reasons unrelated to this chapter, I have focused on research on two kinds of behavior—aggression and sexual behavior—reviewing the kinds of empirical studies being done and then looking at the theoretical and polemical writing to provide a guide to their broader intellectual contexts. Thus, examples of research that I cite here are drawn from this material.

2. Burgess and Molenaar (1993).

3. See Longino (2001) for discussion of the variety of dimensions of such behavior.

4. Mason and Frick (1994).

5. Brunner et al. (1993). Five members of the family exhibited extreme levels of violence, whereas nine others exhibited more moderate, but still higher levels of violence.

6. Hamer et al. (1993).

7. McGue (1994); Scarr (1992, 1993); Plomin, Owen, and McGuffin (1994).

8. A recent paper (Anderson and Bushman, 2002) identifies five different theoretical orientations falling under this general designation.

9. Lavigueur, Tremblay, and Saucier (1995).

10. Adam (1996). For other anthropological work, see Ortner and Whitehead (1981) and Herdt and Stoller (1990).

11. Haapasalo and Tremblay (1994).

12. Bierman and Smoot (1991).

13. Baumrind (1993).

14. Gottlieb (1991). Susan Oyama is one of the principal theorists and spokespersons for the developmental systems approach. See Oyama (1985, 2000).

15. Eldredge and Grene (1992).

16. See Thelen (2000) and Thelen and Smith (1994).

17. Lerner (1991).

18. Cf. Johnson (1993); Edelman, Gall, and Cowan (1984).

19. Ferris (1993, 1994).

20. For an early study see Biegon, Segal, and Samuel (1979).

21. Coccaro, Gabriel, and Siever (1990).

22. Similar issues to those studied by Rasmussen (1995) in the development of electron microscopy will undoubtedly arise with the future development of the various visualizing technologies.

23. Raine et al. (1994).

24. LeVay (1991).

25. For example, some studies such as Heiligenstein et al. (1992) received funding from the Eli Lilly pharmaceutical company.

26. For a preliminary discussion of problems in the definition of behaviors, see Longino (2001, 2002).

27. For a good discussion of what genes can be said to cause or determine, and how, see Waters (2000).

28. Byne and Parsons (1993).

29. Wahlsten and Gottlieb (1997), p. 172.

30. While behavior geneticists don't observe this refinement in practice, it is not clear that the reaction range concept is not capable of this subtlety. See Turkheimer, Goldsmith, and Gottesman (1995). As reaction-norm and reaction range are used, they highlight different aspects of the complex gene–environment relationship.

31. See, for example, the discussion sparked by recent books by Pinker (2002) and Riddley (2003).

REFERENCES

Adam, Barry (1996) "Age, Structure, and Sexuality: Reflections on the Anthropological Evidence on Homosexual Relations." *Journal of Homosexuality* 11, no. 34: 19–33.

Anderson, Craig A., and Brad J. Bushman (2002) "Human Aggression." *Annual Review of Psychology* 52: 27–52.

Baumrind, Diana (1993) "The Average Expectable Environment Is Not Good Enough: A Response to Scarr." *Child Development* 64: 1299–1317.

Biegon, A., M. Segal, and D. Samuel (1979) "Serotonin Activity in Rhesus Behavior." *Psychopharmacology* 61: 77–81.

Bierman, Karen, and David Smoot (1991) "Linking Family Characteristics with Poor Peer Relations: The Mediating Role of Conduct Problems." *Journal of Abnormal Child Psychology* 19, no. 3: 341–56.

Brunner, G. H., et al. (1993) "Abnormal Behavior Associated with a Point Mutation in the Structural Gene for Monoamine Oxidase A." *Science* 262, no. 5133: 578–80.

Burgess, Robert C., and Peter Molenaar (1993) "Human Behavioral Biology." *Human Development* 36: 36–54.

Byne, William, and Bruce Parsons (1993) "Human Sexual Orientation: The Biologic Theories Reappraised." *Archives of General Psychiatry* 50, no. 3: 228–39.

Coccaro, Emil, Steven Gabriel, and Larry Siever (1990) "Buspirone Challenge: Preliminary Evidence for a Role for Central 5-HT-1a Receptor Function in Impulsive Aggressive Behavior in Humans." *Psychopharmacology Bulletin* 26, no. 3: 393–405.

Edelman, Gerald, W. E. Gall, and W. M. Cowan, eds. (1984) *Molecular Bases of Neural Development*. New York: Wiley and Sons.

Eldredge, N., and M. Grene (1992) *Interactions: The Biological Context of Social Systems*. New York: Columbia University Press.

Ferris, Craig, et al. (1993) "An Iodinated Vasopressin (V1) Antagonist Blocks Flank Marking and Selectively Labels Neural Binding Sites in Golden Hamsters." *Physiology & Behavior* 54: 73–47.

Ferris, Craig, et al. (1994) "Septo-Hypothalamic Organization of a Stereotyped Behavior Controlled by Vasopressin in Golden Hamsters." *Physiology & Behavior* 55, no. 4: 755–59.

Ferveur, Jean-Francois, et al. (1995) "Genetic Feminization of Brain Structures and Changed Sexual Orientation in Male *Drosophila*." *Science* 267: 902–5.

Gottlieb, Gilbert (1991) "Experimental Canalization of Behavioral Development: Theory." *Developmental Psychology* 27, no. 1: 4–13.

Haapasalo, Jaana and Richard Tremblay (1994) "Physically Aggressive Boys From Ages 6 to 12: Family Background, Parenting Behavior, and Prediction of Delinquency." *Journal of Consulting and Clinical Psychology* 62, no. 5: 1044–52.

Hall, Jeffery (1994) "The Mating of a Fly." *Science* 264, no. 5166: 1702–14.

Hamer, Dean, et al. (1993) "A Linkage Between DNA Markers on the X Chromosome and Male Sexual Orientation." *American Journal of Human Genetics* 53, no. 4: 844–52.

Heiligenstein, John, et al. (1992) "Fluoxetine Not Associated with Increased Violence or Aggression in Controlled Clinical Trials." *Annals of Clinical Psychiatry* 4, no. 4: 285–95.

Herdt, G. H., and R. Stoller (1990) *Intimate Communications: Erotics and the Study of Culture*. New York: Cambridge University Press.

Johnson, Mark H., ed. (1993) *Brain Development and Cognition*. Oxford, UK: Basil Blackwell.

Lavigueur, Suzanne, Richard Tremblay, and Jean-François Saucier (1995) "Interactional Processes in Families with Disruptive Boys: Patterns of Direct and Indirect Influence." *Journal of Abnormal Child Psychology* 23, no. 3: 359–78.

Lerner, Richard (1991) "Changing Organism-Context Relations as the Basic Process of Development: A Developmental Contextual Perspective." *Developmental Psychology* 27: 27–32.

LeVay, Simon (1991) "A Difference in Hypothalamic Structure between Heterosexual and Homosexual Men." *Science* 253, no. 5023: 1034–37.

Longino, Helen (2001) "What Do We Measure When We Measure Aggression?" *Studies in History and Philosophy of Science* 32, no. 4: 685–704.

Longino, Helen (2002) "Behavior as Affliction: Framing Assumptions in Behavior Genetics." In *Mutating Concepts, Evolving Disciplines: Genetics, Medicine, and Society*, ed. Rachel Ankeny and Lisa Parker. Boston: Kluwer Publishing.

Mason, Dehryl, and Paul Frick (1994) "The Heritability of Antisocial Behavior: A Meta-Analysis of Twin and Adoption Studies." *Journal of Psychopathology and Behavioral Assessment* 16, no. 4: 301–23.

McGue, Matt (1994) "Why Developmental Psychology Should Find Room for Behavioral Genetics." In *Threats To Optimal Development: Integrating Biological, Psychological, and Social Risk Factors*, ed. Charles Nelson Alexander, et al., pp. 105–19. Hove, England: Lawrence Erlbaum Associates, Inc.

McGue, Matt, Steven Bacon, and David Lykken (1993) "Personality Stability and Change in Early Adulthood: A Behavioral Genetic Analysis." *Developmental Psychology* 29, 1: 96–109.

Ortner, Sherry, and Harriet Whitehead (1981) "Introduction: Accounting for Sexual Meanings." In *Sexual Meanings*, ed. Sherry Ortner and Harriet Whitehead, pp. 1–27. Cambridge: Cambridge University Press.

Oyama, Susan (1985) *The Ontogeny of Information*. New York: Cambridge University Press.

Oyama, Susan (2000) *Evolution's Eye: A Systems View of the Biology-Culture Divide*. Durham, NC: Duke University Press.

Pinker, Steven (2002) *The Blank Slate*. New York: Viking.

Plomin, Robert, Michael Owen, and Peter McGuffin (1994). "The Genetic Basis of Complex Human Behavior." *Science* 264, no. 5166: 1733–39.

Raine, Adrian, et al. (1994) "Selective Reductions in Prefrontal Glucose Metabolism in Murders." *Biological Psychiatry* 36: 365–73.

Rasmussen, Nicholas (1995) "Mitochondrial Structure and the Practice of Cell Biology in the 1950s." *Journal of the History of Biology* 28: 381–429.

Riddley, Matt (2003) *Nature via Nurture*. New York: HarperCollins.

Scarr, Sandra (1992) "Developmental Theories for the 1990s: Development and Individual Differences." *Child Development* 63, no. 1: 1–19.

Scarr, Sandra (1993) "Biological and Cultural Diversity: The Legacy of Darwin for Development." *Child Development* 64: 1333–53.

Thelen, Esther (2000) "Grounded in the World: Developmental Origins of the Embodied Mind" *Infancy* 1,1: 3–28.

Thelen, Esther, and L. B. Smith (1994) *A Dynamic Systems Approach to the Development of Perception and Action*. Cambridge, MA: MIT Press.

Turkheimer, Eric, H. Hill Goldsmith, and Irving Gottesman (1995) "Some Conceptual Deficiencies in 'Developmental' Behavior Genetics: Comment." *Human Development* 38, no. 3: 143–53.

Wahlsten, Douglas, and Gilbert Gottlieb (1997) "The Invalid Separation of Effects of Nature and Nuture: Lessons from Animal Experimentation." In *Intelligence, Heredity, and Environment*, ed. Robert Sternberg, Elena Grigorenko, et al., pp. 163–92. Cambridge: Cambridge University Press.

Waters, C. Kenneth (2000) "Molecules Made Biological." *Revue Internationale de Philosophie* 4, no. 214: 539–64.

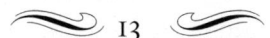

13

INTROSPECTIVE EVIDENCE IN PSYCHOLOGY

Gary Hatfield

Introspection was once the mainstay of psychological research, the primary source of psychological evidence. But, as history has it (e.g., Lyons, 1986, chs. 1–2), in the first part of the twentieth century introspection was discredited by behaviorists in psychology, and by the likes of Wittgenstein and Ryle in philosophy. These critics purportedly showed that introspection was unscientific, conceptually impossible, or akin to believing in ghosts. As a result, introspection disappeared as a source of evidence in psychology and philosophy alike (Lyons, 1986).

This standard account, as most do, contains a grain of truth. Introspection, broadly conceived, was once the primary source of evidence in experimental psychology—although it was never, in any period of psychology's long past, considered to be the only source of evidence (Hatfield, 2003b; Titchener, 1912a). Even as one among several sources of evidence, introspection was in decline by the middle of the twentieth century, both in philosophy and psychology, largely because of attack from the behaviorists. Although the use of introspective evidence was not fully abandoned, criticisms from within psychology put an end to *analytical introspection,* narrowly defined to mean a specific method of seeking the "atomic" elements of experience.

Interest in introspection has recently revived, in two contexts. In connection with questions about first-person knowledge, some authors have offered positive accounts of self-knowledge of some mental states, especially opinions and convictions (e.g., Moran, 2001). In connection with theories of consciousness, the question of introspective access to the self, or to private conscious states, has drawn considerable attention (e.g., Armstrong, 1980; Lycan, 1997). All the same, the notion of introspective awareness of specifically phenomenal aspects of perceptual states remains deeply suspect (Dretske, 1995, ch. 3; Jackson, 1998, ch. 4; Tye, 1995, ch. 5).

I am a friend of introspection; I introspect regularly. I think I find things out—though not everything there is to know, even about my own mind. I turn to introspection frequently in thinking about perceptual experience, and in testing claims made by perceptual psychologists. More importantly, I believe that introspection maintains an ineliminable role in psychology itself, as a source of evidence. This is especially apparent in perceptual psychology, which will be my ultimate focus.

260

In preparation for examining the place of introspective evidence in scientific psychology, I begin by clarifying what introspection has been supposed to show, and why some concluded that it couldn't deliver. This requires a brief excursus into the various uses to which introspection was supposed to have been put by philosophers and psychologists in the modern period, together with a summary of objections. I then reconstruct what I take to have been some of the actual uses of introspection (or related techniques, differently monikered) in the early days of experimental psychology. Here, I distinguish broader and narrower conceptions of introspection, and argue that recent critics have tended to misdescribe how introspection was supposed to work. Drawing upon the broader conception of introspection, I argue that introspective reports are ineliminable in perceptual psychology. I conclude with some examples of such ineliminable uses of introspective reports in both earlier and recent perceptual psychology.

INTROSPECTIVE OBJECTIVES

Introspection, broadly conceived, describes a mental state or activity in or through which persons are aware of properties or aspects of their own conscious experience. Being aware that one feels cold, is seeing red, or is worried, if mediated by awareness of conscious experiences that include feeling cold, seeing red, or being worried, are all instances of introspection. This broad description (which I refine in the text that follows) is intended to cover the variety of uses ascribed to introspection in the history of philosophy and psychology.

Introspection has been undertaken with the aims of both self-knowledge and knowledge of the self or mind. Self-knowledge, the use of introspection most typically discussed now by philosophers (e.g., Armstrong, 1980; Moran, 2001; Myers, 1986; Shoemaker, 1996), is knowledge of what is peculiar to a given person. This may include their beliefs and memories, and also may describe allegedly private or wholly subjective states of consciousness (experiences of sense-data were supposed to be such). By contrast, the search for knowledge of the self or mind is undertaken with the aim of attaining general, or intersubjectively common, descriptions of the self or mind. This general knowledge is to be achieved via introspective observations and their report. Psychophysics, in which subjects might match stimuli according to their appearances in specified circumstances, is an example. In such cases, introspection is supposed to serve as a basis for generalizations about all (or most) human selves or minds.

With this distinction between differing aims for introspection in place, let us consider the most important objectives for introspection, actual or purported, in the history of philosophy and psychology.

Explicit appeal to introspection is found in Augustine and Aquinas, and such appeals became widespread and prominent in the seventeenth century (Lyons, 1986, ch. 1). Descartes especially is linked with the early history of introspection. His *Meditations* contain a studied turning away from the body, a "looking within" to find the foundations of knowledge. Purportedly, he discovered these foundations in incorrigibly known states of mind, from which he sought to infer the properties of a world beyond the mind.

What did Descartes claim to find when he turned inward? Opinions vary. Later philosophers, including Hume and Kant, argue as though Descartes or his rationalist descendants claimed to perceive the soul as a simple substance, by a kind of direct inspection. Such perception of the soul or mind is our first purported objective for introspection:

(1) To perceive the mind as a simple substance.

Hume and Kant did not describe what they believed philosophers such as Descartes held this perception to be like; they merely asserted that *they* did not find a simple soul manifest in their inner experience.[1] One might assume that they believed Descartes and others had claimed to "see" a punctiform entity, a speck of immaterial substance.[2]

Although there was talk in the early modern period of whether the soul should be regarded as a point, Descartes refused to attribute to the soul any predicates derived from extension (1641/1984, 266). More importantly, he never claimed to perceive the soul itself directly, as a simple substance. Rather, he claimed to perceive important features of the "nature of mind" via reflection.[3] According to Descartes, the mind manifests various characteristic types of experiences and various types of mental activity in relation to those experiences, which include perceiving through the senses, making judgments about such perceptions, imagining, remembering, and understanding, or willing various things, feeling bodily sensations such as hunger and pain, and undergoing various passions or emotions—or at least seeming to do, feel, or undergo these acts and experiences (1641/1984, 19). From further reflection on and conceptualization of these mental activities, Descartes arrived at some conclusions about the nature of the human mind: that it is essentially an immaterial substance; that intellect and will are the two basic faculties of mind; and that mind is distinct from, but interacts with, the human body (1644/1985, 204, 208–19).

We thus have another objective for introspection:

(2) To discern the nature of mind.

This objective might be intended to rest upon the sort of intellectual perception (or intuiting) of the essence of mind that Descartes claimed to achieve: not "seeing" a speck, but understanding an essence. Alternatively, this objective might arise from the aim of knowing what a mind is by describing what it does, that is, by cataloging various mental activities. This more specific objective is:

(3) To discern the characteristic states and activities of mind.

An example of such a characterization is the claim (made prominent in the eighteenth century) that the three main divisions of mental life are perceiving, feeling, and desiring, rather than (as Descartes had it) perceiving and willing only.

Those investigating the mind by reflecting on their experience might observe that they can know more particular qualitative and temporal features of their mental states, or at least of those available in consciousness. Such features might include the division of sensory perceptions into various quality groups, or modalities, such as vision, touch, hearing, taste, and smell. Such investigators might claim to compare the intensities or durations of various sensations, feelings, desires, and thoughts. This objective is:

(4) To ascertain the qualitative features and temporal relations of conscious states.

Such claims can be found in Descartes, but also in Hume, the mental geographer (1748/1999, 93). In the latter part of the nineteenth century, experimental psychology took as two of its principal aims (a) characterizing quality groups through the experimental techniques of psychophysics, and (b) measuring temporal relations in mental processes.

One might hope, from observations of such qualitative features and temporal relations, to discover or infer the basic psychological processes or operations of the mind. We thus have a more specific version of (3), which is our fifth aim:

(5) To discover or infer the character of mental or psychological processes.

This aim was vigorously pursued in various theoretical contexts in the late nineteenth and early twentieth centuries. On the basis of observations, various psychologists claimed to discover characteristics of psychological processes:

263

(5.1) That they involve pure acts of intellect or imageless conceivings

(5.2) That they are always imagistic

(5.3) That such processes manifest genuine activity, as in attending or willing

(5.4) That some processes are unconscious.

Findings (5.1) and (5.2) report the opposing views in the imageless thought controversy that raged in the early twentieth century (see Kusch, 1999). Some psychologists claimed to discover a phenomenology of thought processes in the absence of any mental images. Others claimed that thought is always directed toward or involves images. During the same period, psychologists disagreed over whether instances of genuine psychological activity are found introspectively (5.3), or whether we in fact have available only experiences that are not direct manifestations of activity, even though the language of activity is used to label them (see James, 1912). Many psychologists in the nineteenth century (and before) used introspective evidence as a basis for positing unconscious (or perhaps unnoticed) psychological processes or operations (5.4) that yield conscious experience, whereas others sought to rule out such processes (see Hatfield, 1990, chs. 2–5; 2003b). These various areas of disagreement fueled the fires of behaviorists and other enemies of introspection.

Uses (1) through (4) are found in earlier philosophical and psychological writings. Uses collected under (5) were taken up by the new experimental psychology in the nineteenth century. Use (4) and some parts of (5) are, or should be, of interest in philosophy and psychology today.

Philosophers interested in epistemology also have made claims for the power of introspection (or "immediate perception") of perceptual data. In the first half of the twentieth century, sense-data were said to be immediately perceived objects of perception, perhaps incorrigibly known, and in any case were the basis for all other empirical knowledge. We thus have a sixth use:

(6) To perceive (incorrigibly known?) sense-data, as the foundation for other knowledge.

Such data were supposed to be private, and to provide an initial basis from which to construct or infer the external world and other minds. Bertrand Russell (1919) developed this talk of "construction," which led him to a position of "neutral monism" that he shared with Ernst Mach and William James (Hatfield, 2004). On this view, only sense-data (or rather "momentary particulars") are affirmed as existing; external objects and minds (including one's self) are regarded as constructions from such data. Adherents of a "representative" theory of perception treat private sense-data as the epistemic basis for knowledge of really existent external objects and other minds (Broad, 1923, pt. 2).

Finally, introspection has been taken to provide individual knowledge about the self. To the extent that Augustine's *Confessions* are seen as the report of a personal spiritual journey, they contain introspective reports of Augustine's personal experiences and reactions. This gives us a seventh use:

(7) To know the particularities of one's self (hopes, aspirations, beliefs).

This is introspection as affording self-knowledge, that is, as providing privileged access to specifically first-person facts. The extent to which this source of knowledge can provide full insight into one's beliefs and desires has long been questioned. But the notion that at least some specifically first-person knowledge is available retains many advocates.

Of these uses, the first five purport to provide evidence for claims about mind, or mental states and processes, in general. They are not intended to provide special knowledge of an individual's own thoughts and beliefs, but are instead aimed at what I term "knowledge of self or mind": knowledge of the characteristics of the mind, or of mental processes, in general. In this context, individual introspective observations are taken to reveal characteristics common to all minds. The claims made on behalf of such uses range from knowledge of the nature of mind, as in (2), to knowledge of its states and processes, as in (3) through (5).

Uses (6) and (7) focus on specifically individual knowledge. Use (7) in particular describes the sort of introspection (broadly conceived) that has been defended of late, under the title "first-person knowledge," by moral psychologists such as Moran (2001) and epistemologists such as Shoemaker (1996). They observe that, in many such cases, we attain self-knowledge by deliberating about what we hope or believe. Because they are *our* deliberations, we attain a specifically first-person knowledge of our beliefs, in the very act of deciding what they are. Moran (2001: 11–20) distinguishes this sort of first-person access from introspective knowledge of the phenomenal aspects of perceptual experience, which he thinks has been discredited. He associates the latter sort of "introspection" with a "perceptual model" of first-person knowledge that allegedly is directed toward a special inner object.

Although uses (6) and (7) fit the broad conception of introspection described above (since they involve conscious awareness of an allegedly private perceptual object, or of one's convictions), they are not my focus here. My primary concern is with reports or responses that serve as evidence for the characteristics of perceptual experience. Such reports or responses were conceived by earlier advocates—and should, I will argue, be so conceived today—as providing intersubjectively valid observational knowledge of (at least some of) the characteristics of perceptual states (and other sorts of mental states). The remainder of this chapter therefore leaves aside the specifically personal knowledge that has been the focus of some recent philosophers, and considers introspection as a source of evidence for general statements about mind, ultimately focusing on its use in providing scientific evidence in the study of perception. In considering objections to introspection in the next section, I therefore focus on uses (1) through (5); ultimately, I seek to vindicate aspects of uses (4) and (5).

OBJECTIONS TO INTROSPECTIVE EVIDENCE

Many objections have been raised against actual or alleged uses of introspective evidence. Some of these objections are powerful and on target, while others have been based on a misdescription or caricature of introspective evidence or its objectives. I want to consider some telling objections, which limit the scope of introspective knowledge, and some misdirected objections, which I hope can be put aside.

Against use (1), which seeks direct introspective awareness of the soul or self as a simple substance, Hume's and Kant's reflections, as described earlier, are persuasive. However, alleged direct phenomenal acquaintance with the soul as a substance was not a mainstream position among metaphysicians of the soul (e.g., Descartes or Leibniz). Rather, they used arguments (which might include phenomenally based premises), rather than direct inspection, to arrive at their conclusions. In any case, it should be granted that introspection does not directly reveal the soul as a simple substance.

The more plausible claim of Descartes and others was that introspection revealed the nature of mind by revealing its characteristic states and activities as in use (2). In this use, Descartes and others could allow that the mind is always perceived through its properties (acts and states); perception of these properties allows one to grasp the mind's essence or nature. I suppose that few philosophers today believe that the nature of mind can be discerned in this way, in part because there are now few philosophical dualists who believe mind to be a separate substance with its own nature or essence. But even taking "nature of mind" more broadly, to include functionally defined characteristics that have been popular of late (e.g., the mind is constituted of symbol-crunching processes), few to none would believe that this nature can be discovered directly through introspection.

Use (3) aims at discerning "the characteristic states and activities of mind." This objective supposes that such characteristic states are accessible to consciousness. Yet it is widely accepted today that many cognitive processes are not accessible to consciousness. One needn't endorse the Freudian unconscious to make the point. Cognitive psychology observes that many processes underlying perception and cognition—from simple visual capacities such as stereoscopic vision, in which minute spatial differences between the two retinal images are compared, to the recognition of a friend by her appearance—take place outside consciousness. Although the *results* may be available to consciousness (in the experience of depth, or in the conscious recognition), the processes are not manifest. Acceptance of the point that some mental states and processes are not present to consciousness would not preclude introspection as a source of evidence, but it would limit its scope.

Three further objections seek to rule out uses (4) and (5). One urges that introspection is unreliable; a second casts aspersion on its object; the third proclaims it to be conceptually impossible. The charge of unreliability was prominent in J. B. Watson's arguments against introspective psychology. He pointed to several examples, including the imageless thought controversy, disagreements over the number of "degrees" of attention, and disagreements over the number of elemental sensations (Watson, 1914: 6–7). These charges seek to undermine the use of introspection, in uses (4) and (5), for determining the character of mental or psychological processes, and the fundamental elements entering into them, by observing the qualitative features and temporal relations of conscious states.

If the aim of psychological introspection is to find the "least elements" of mind, or to fully reveal the fundamental acts of mind (as in levels of attention), then Watson's charges stand. However, the problem may not be with introspection itself, but with the theoretical framework in which it was used. The notion that there are least elements of sensation to be discerned introspectively is, as James and the Gestalt psychologists (among others) observed, a theoretical construct. No one ever experiences a bare least element of sensation; such elements are posited by theory (see Hatfield, 1990, chs. 4–5). If the theory is wrong, then the sought-for least elements will not be discovered. As for the notion of levels of attention, it may have been an attempt to attribute overly fine phenomenological distinctions to the dynamics of attention. In a wider context, introspective techniques (broadly construed) are still used in studying attention, and to good effect (as discussed below).

267

Finally, the imageless thought controversy attempted to use introspection to discern facts about the content and structure of higher cognitive processes. The techniques were in many ways dissimilar from those used to elicit introspective reports in perception, because the object of experience was less restricted and defined. Such methods, which were sometimes called "systematic experimental introspection," came in for strong criticism from other introspective experimentalists (as reviewed by E. B. Titchener, 1912b). Presumably, the adjective "systematic" was supposed to promote the legitimacy of the techniques in question, in which subjects might simply be asked to record observations on their thought processes after having carried out a given cognitive task. We may agree with Titchener's (1912b) conclusion that this form of introspection is of dubious reliability. However, the question of the reliability of introspective techniques in other contexts remains open.

The remaining two objections seek to rule out the possibility of introspection by casting aspersion on its objects or by claiming that it is conceptually impossible. These objections start from the assumption that to characterize introspection as "inner perception" or "inner observation" is to presuppose a special inner object. In the case of sense perception, which is my focus in this chapter, this object is supposed to be an existent thing that is distinct not only from external objects but also from perceptual experience of external objects. The existence, or the knowability of such inner objects is then challenged on various metaphysical and epistemological grounds.

Metaphysically, it has been suggested that experiential states containing their own mental or subjective content should not be countenanced, because to do so would be like believing in ghosts or other "unnatural" entities. Watson (1914: 20) and B. F. Skinner (1963) championed this sort of claim, and it can be found in many recent philosophers who would banish all mentalistic notions that cannot be "naturalized." Such philosophers may hold that notions such as "information" and "representation" can be naturalized by employing an engineering conception of "information" that rests on natural relations (such as conditional probabilities) among properties or states of affairs. Hence, the notion of representation may be retained, but, they argue, qualitative experiential states would be spooky entities that don't fit into a naturalistic outlook (see, e.g., Rey, 1997: 255, 301).

This argument is directed against the alleged object of introspective awareness. It denies that introspection can provide a distinctive source of knowledge of the phenomenal, on the grounds that in order to do so, introspection would have to be directed upon a phenomenal object of dubious metaphysical status. It seeks to restrict the sort of evidence that psychologists should countenance, on the basis of a metaphysical assertion about what is natural and what is not. And yet the basis for the claim of what does or does not "belong to nature" is not spelled out. With the decline of philosophers' presumptions to have distinctive *a priori* insights into the fundamental elements of nature, such claims must in some way make contact with empirical knowledge. One typical way to decide on the range of natural states, processes, and objects is to look to the generalizations of the natural sciences. But if psychology is included among the natural sciences, the question of whether introspectible experiential states are found in nature reduces to the question of whether they are the object of generalizations (or other scientific assertions) in psychology. Philosophers' intuitions about what is natural and what is not would in this case give way to the question of what is being (or can be) studied in, and what is posited by, perceptual psychology.

Turning to the epistemological objection, introspection is supposed to be conceptually impossible because it would require inward-looking descriptions of a sort that will not bear scrutiny. Wittgenstein's "private language argument" is supposed to tell against such descriptions. As the story goes,[4] the language in which we describe our perceptual experience is parasitic upon, and perhaps presupposes the existence of, external objects. As it was sometimes put, "is red" is more fundamental than "looks red." This means that our descriptions always start from attributions of properties to publicly perceivable things. From this, it is concluded that there is no conceptual space for knowledge or description of what is private or internal. Such descriptions would be forced to employ concepts proper to external objects, which is incompatible with the purportedly private status of their objects. Hence, the old idea that we start from awareness of inner states, so described, and work out to awareness, belief, or knowledge of an external world is considered to be conceptually undermined. On this argument, uses (4) through (6) are ruled out.

Any reply to this argument must distinguish various purposes one might have in attempting to describe one's phenomenal experience. If one proceeds from the traditional philosophical aim of describing the foundation or basis for knowledge of the external world, then, if that foundation or basis is supposed to be conceptually independent of beliefs or knowledge about an external world, the "private language" argument could have some bite. But if the aim of describing or attending to phenomenal experience is simply to discover how things look, in the sense of how the world is perceptually presented to observers under specified conditions, then the matter is not so clear. In these circumstances, I might use the descriptive language applied to external objects in order to direct attention to aspects of how those objects are experienced. This is the case in contemporary perceptual psychology, which supposes that one can describe one's own experience using terms that are also used for describing the properties of external objects.

Let us take an example from color perception. In asking for reports on the colors of things, psychologists may instruct subjects to distinguish between the color they take an object actually to have, and the way in which the object looks or appears.[5] If, as I examine a piece of paper on my desk, I am asked what color the paper looks to be, I may unhesitatingly say that it looks white. I perceive it to be white. An experimenter might then ask me to attend carefully to how the paper looks, and to respond to whether it appears with the same whiteness all over. Noting my uncertainty about the task, the experimenter may explain that I am to distinguish the question of whether I would judge or estimate the paper to be uniformly white (as opposed to being dyed or otherwise colored at any place), from my report on its current *appearance* as regards sameness or variation of color. Under such instructions, I report (let us imagine) that although I certainly judge the paper to be the same white everywhere, and I do so because of how it looks, nevertheless the paper appears darker and lighter in different areas across its surface, and it has a reddish tinge in one portion. Warming to the task, if I were next asked to go beyond simply describing the paper's appearance so as to explain why it looks this way, I would say that because of its slight curvature and the direction of the light, the paper appears darker and lighter across its surface, and that it is next to a red ceramic cup that has reflected some reddish light onto it. I would still be in no doubt that the paper *is the same white* across its entire surface. I would not say that it looks to me as if it is a piece of paper that is white in some areas but has been colored darker (or grey) in another area, and red in yet another. It looks to me as uniformly white paper does in many ordinary circumstances. All the same, I am able to use terms such as "white," "grey," and "red" to describe the varying appearance across the surface of the paper (even if I were myself unable to explain those appearances, but simply reported how the paper looked).[6]

In these circumstances, the experimenter has co-opted my ability to use color words to recognize the colors of objects by asking me to describe subtle variations in appearance. If I were claiming that the descriptions of the appearance were epistemically primary and provided the conceptual basis for constructing my knowledge of the external world, I might be in trouble. But if the aim is simply to describe in detail how the paper looks—where I've distinguished the attempt to describe the paper's appearance from what I would conclude about how the paper is physically constituted—then this problem about what is conceptually primary does not arise. I may develop and elaborate concepts of the phenomenal using whatever materials are available, including predicates normally used to describe external objects.

There is, however, another objection raised against "inner" description, or describing the looks of things. It is a phenomenological objection, based upon a report of how things look. According to this objection, phenomenally our experience seems to be "out there," not "in here." But introspection is supposed to be a "looking within," and is supposed to take as its object something besides the external object.

This objection has been stated by many authors (e.g., Dretske, 1995: 54, 62; Harman, 1990; Tye, 1995: 30). Georges Rey sums it up as follows: "as a number of writers (e.g., Harman, 1990; Dretske, 1995) have stressed, a great deal of what passes for introspection of one's 'inner' experience consists of reports about how the *outer* world seems: we don't so much report on the features of the 'inner movie' as upon what that movie *represents* (e.g., that *barns* seem red, *the sky* a dome)" (Rey, 1997: 136–37). That is, we don't seem phenomenally to be attending to a special inner object. Rather, when asked to report, say, on the color of a piece of paper, it seems to us that all we see is a piece of paper.

Rey (1997: 136) describes this as a "problem" for introspection. How so? In fact, two points are compressed together here. Harman (1990), Dretske (1995), and Tye (1995) all wish to reduce qualitative content in perception to the bare representation of properties of external objects. Hence, they deny that in perceptual experience we are presented with qualities that arise from how we subjectively represent objects (what Tye [2002] calls "qualities of experience"), such as is thought to be the case if we treat color as a subjective quality that serves as a mere "sign" for its cause in the object (Hatfield, 2003c). They don't want there to be features of our experience that depend on the subject's way of representing things, rather than on external-world content about objects. This sort of point cannot, of course, be decided just by reporting on the "diaphanous" or "transparent" character of our experience (names for the fact that in visual perception we seem to see the external world directly, without anything intervening, particularly not our own mental states); it depends on a substantive account of the meta-physics of perceptual qualities (such as color). But surely a decision on the metaphysics of qualities shouldn't be required before we can decide whether it is possible to describe phenomenal aspects of our perceptual experience. For even if these authors were right about the metaphysics, we could still ask how the external thing looks. Hence, this part of their "transparency" position is not relevant to my inquiry into the use of introspection as a source of evidence in psychology. The question of whether we can report on our experience does not require a prior solution to the metaphysical problem, even if such a solution might influence our view of what there is to report.

271

This brings us to the second point, which concerns what it is like to attend to our own experiences. The phenomenological point about transparency is supposed to undermine a notion of introspection as describing "inner" experience.[7] Our experiences seem "transparently" to be of external things; we don't seem to be aware of some inner object. But introspection is supposed to be "inner." Hence, at least in the case of sense perception, introspection does not find its intended object and so can be dismissed.

This objection is founded upon a misconstrual, or caricature, of how introspection has long been supposed to work. If we distinguish (a) the metaphysical question of whether introspection is directed upon objects that are distinct from external objects (as "sense-data" are posited to be, or as phenomenal qualities might be), from (b) the phenomenal locating of the objects of introspection, we will find that very few authors in the history of psychology or philosophy held that sense-perceptions are experienced as "inner." The early experimental psychologists who advocated introspection certainly did not. The relevant question then is not whether our experience seems to be "in here" or "out there," but whether any relevant differences exist between simply observing external objects and observing the experiences we have in doing so. Classical experimental psychology held that such differences exist. This point requires elaboration.

ACTUAL PRACTICES OF INTROSPECTION IN PSYCHOLOGY

The notion of introspection was refined over the course of the nineteenth century, partly in response to various charges that introspective observation is impossible. Comte (1830–42/1855: 33) argued that direct introspection of mental processes is not possible, because it would interrupt itself. The initial response to this charge was to grant that, although any attempt to observe our own thought processes directly would interrupt itself, we can "observe" by seeking to remember our thought processes just after they have taken place (J. S. Mill, 1865: 64); introspection could operate via memory. Franz Brentano

(1874/1995: 29–36) refined this response by allowing that introspective *observation* is possible only through memory, but he contended that there is also a kind of "incidental" perception of our mental states while we are having them. He called the fleeting awareness that we are having a certain sort of mental state an "inner perception," distinguishing it from introspection proper, which he called "inner observation."

These discussions took place before the widespread use of introspective techniques in experimental settings, which increased dramatically after 1880. At first, experimental psychologists such as Wilhelm Wundt agreed with Brentano's point that self-observation (*Selbstbeobachtung*) of mental states and processes is unreliable because it interrupts itself (Wundt 1882/1885: 136–37). But he soon reversed himself on this point, and refined his position.

In a lengthy study on introspection, Wundt (1888) distinguished inner perception from self-observation (introspection proper). He characterized self-observation as the "deliberate and immediate observation of inner processes" (1888: 297). The key terms all require explanation.

"Deliberate" implies that the subject directs his or her attention to the states or processes being observed. Although deliberate observation need not be previously planned (a botanist may observe with deliberateness a specimen that she has found serendipitously during a walk), it does suggest that the observer is paying careful attention to the object of observation. Wundt described "scientific observation" as the "deliberate direction of attention to the phenomena" (1888: 293). The observer is prepared to discriminate or discern characteristics of the observed phenomena, and to remember the results of such discrimination. The observer may direct his or her attention to selected aspects of the phenomena. Without this directing of attention, mere "inner perception" may allow us to be aware of the contents of our minds, but not in the deliberate manner of introspection. Introspection proper involves deliberate consideration with the intent of discriminating among psychological states or processes.

"Immediate" rules out the sort of memory-mediated introspection that Mill and Brentano allowed. According to Wundt, observation requires the presence of what is observed. Hence, retrospection does not count as introspection ("self-observation") of the remembered state or process; rather, it is observation (or introspection) of a present memory of the past state or process (1888, 294: 297–300). As such, it need not be wholly untrustworthy, but it introduces the usual limitations on memory as a source of evidence.

According to Wundt, experiment makes deliberate introspection possible (1888: 301–3). Experimental conditions allow the subject or observer to maintain an object of observation over a period of time, as when an observer is asked to match color samples. The samples can be examined for a preset period (two seconds, say), or subjects might simply be asked to declare the match only when they are sure. Also, an experimenter can elicit the same (or closely similar) psychological processes by arranging for an exact repetition of external conditions.[8] If higher thought processes were the object, the immediate results of techniques of directed attention would be suspect if considered to be observations of a constant object, for the directed attention of introspection might interrupt the thought process, and the thoughts themselves might alter from trial to trial as the result of learning or speculation by the observer. But in color matching, observation of properly arranged color stimuli for a few seconds introduces minimal change, and if proper precautions are taken, later observations will not be systematically altered by physiological or psychological after-effects from previous observations.

"Inner" is the main offending term, according to those who emphasize the "transparency" of perception (Dretske, 1995; Harman, 1990). Many philosophers have supposed that this term must imply that the object of introspection seems to be "in the mind" or "in the head," rather than in the world—otherwise, the point about transparency would have no bite. They have also supposed that it implies an ontologically distinct entity, a "sense-datum" or a subject-dependent *quale*. Finally, they have supposed that, failing the existence of such an entity, there would be nothing to do in introspection except report on the external object; there could be no "observation" of one's own experience.[9]

None of these assumptions applies to the classic notion of introspection as experimental psychologists such as Wundt developed it. In describing the objects of introspection, Wundt did not posit an "inner" location or require a specific metaphysical theory of qualia, but he nonetheless did allow an attitude of observation toward one's own experience. According to him, introspection takes the same (phenomenal) objects as ordinary perception, but approaches them with a different attitude or "point of view." Whatever may be the truth about the relation between the physical and the psychical (in sense perception, let us say), the objects of observation are the same. Wundt rejected the definition of psychology as the "science of inner experience," for the reason that "it may give rise to the misunderstanding that psychology has to do with objects totally different from the objects of so-called 'outer experience'" (1901/1902: 2). Various perceptions, as of "a stone, a plant, a tone, a ray of light," can be viewed either as natural phenomena, or as "ideas" or presentations to a subject.

As a psychologist, Wundt was not concerned to determine the metaphysical status of perceptual experience. He regarded sense experience as presenting external objects (4, 13). Methodologically, psychologists can study the same observational phenomena as do physicists, but they do so from a different point of view: "the expressions outer and inner experience do not indicate different objects, but *different points of view* from which we take up the consideration and scientific treatment of a unitary experience" (3).[10] And again: "from this point of view, the question of the relation between psychical and physical objects disappears entirely. They are not different objects at all, but one and the same content of experience" (1901/1902: 11).

In physics and chemistry, one seeks to describe objects by abstracting away from a subjective point of view as much as is possible (1901/1902: 3, 357). Those sciences develop their own concepts, which describe objects in terms of stable physical and chemical properties (e.g., mass, force, acid, base, etc.). Psychology, by contrast, studies all aspects of experience, including those that are momentary, or that depend on momentary relations between the subject and an external object. Even though the objects in introspective studies of perception typically are existing external objects,[11] psychologists have developed special phenomenological concepts for describing them. These include notions such as "sensation," and, for a given sensory modality, the range of sensory qualities, such as hue, brightness, and saturation in the case of color. In a psychological investigation, one may ask subjects to describe the objects of perception phenomenally, in terms of how the objects look from moment to moment.

As an example of the two viewpoints, consider first a student of chemistry in the chemistry lab who simply wants to know whether the litmus paper she has just dipped into a liquid has turned red or blue. She isn't interested in whether the appearance of those colors varies with the variations in the lighting found in the chemistry classroom. Rather, the classification into *red* or *blue* (depending on whether the liquid was an acid or a base) is binary. In the psychology classroom down the hall, a student might simply report the classification into red and blue introspectively, as his awareness that the paper looked red or looked blue.

But that bare classification into color classes need not exhaust the experimenter's interest in the look of the colored papers. In a study of color perception, a subject might be shown colored papers matching the red and blue of litmus paper under various conditions of illumination, and be asked to compare how they look (their appearances). Far from simply declaring "red" or "blue," the student might note that (under ordinary illumination) the "red" sample would more accurately be described as pink, and the "blue" sample as bluish-violet. He might also note that the samples take on differing phenomenal hues under variations in lighting, that the two swatches nonetheless continue to be distinguishable in color, and that under a wide range of illuminations he could still easily classify the samples as "red" or "blue," even though each sample doesn't look exactly the same under all those conditions.

In describing how the litmus paper looks during the experiment, the psychology student employs concepts of phenomenal appearance rather than binary color classification. He never once need conceive of himself as accessing a special inner object that seems to be located within himself: he is always describing how the colored papers *look*.[12] Hence, whether the colors he reports are in fact subject- or perceiver-dependent (a metaphysical question we have, for now, put aside), the traditional object of introspection in the study of visual perception is characterized as *phenomenally* outer.

Wundtian experimentation took place in highly controlled conditions and used trained observers. When Wundt and others coupled this experimental practice with certain further theoretical assumptions, such as that sensory experience is constituted out of punctiform sensational "elements" or "atomic sensations," they evolved the introspective practices that are classed under the name of "analytical introspection."[13] In a broad sense, analytical introspection simply means introspection undertaken in order to discriminate and classify experiences (Titchener, 1912b: 495–96). Here, "analytical" means classificatory, and the notion is unobjectionable. But in the narrow sense, it means introspection undertaken to uncover atomic sensations (1912b: 495). Here, "analytical" means resolution into least elements; in vision, these elements were (by hypothesis) punctiform sensations. In this latter guise, analytical introspection came in for heavy criticism from James (1890, ch. 6) and the Gestalt psychologists.[14]

Wolfgang Köhler devoted a chapter of his *Gestalt Psychology* (1947) to criticizing this form of introspection. He was especially concerned to question the notion that "hidden" elements, called "pure sensations," underlie phenomenal experience as we have it. The Gestalt psychologists emphasized that ordinary experience is of a world at a distance, experienced in three dimensions. They held that objects nearby are ordinarily experienced under conditions of spatial "constancy." This means that a dinner plate seen across the table (and so, at an angle of 45° to the line of sight) is nonetheless perceived as circular, rather than as an ellipse (its projective shape on the retina).[15] By contrast, an analytical introspectivist might hold that the "real" sensation conforms to the two-dimensional projection, while the fact (or—depending on the particularities of the theory—the report) that the plate looks round would be ascribed to learning.

The Gestalt psychologists held that an accurate phenomenal report of how the plate looks would say that it looks *round*. They would ascribe the perception of it as an ellipse to circumstances in which the observer has adopted a special attitude, sometimes called the "painter's attitude," such as one might learn in drawing class when attempting to produce a perspectival picture of the plate. According to Köhler and his colleagues, such deliberately elicited experiences should not constitute the starting point in the psychology of perception. Perceptual psychology, or indeed any psychology, must start from the external world "as we have it": "There seems to be a single starting point for psychology, exactly as for all the other sciences: the world as we have it, naively and uncritically. The naivete may be lost as we proceed" (1947: 3). He called the world as we have it the world of "direct experience." And he contended that such experience is "the raw material of both physics and psychology" (1947: 34).

Although rejecting analytical introspection, the Gestalt psychologists nonetheless relied on introspection, more broadly conceived. In their writings, they use demonstration drawings as a means of making readers become aware of aspects of their perceptual experience. The drawings illustrated figure/ground relations, grouping of phenomena, and other Gestalt principles. A familiar example is the Necker cube, which is like a drawing of a wire cube seen from one of its faces. The cube reverses in depth as one looks at it. One experiences this reversal, and can attend to it and report on it. Such reports, although not labeled introspective by Köhler (who limited that term to analytical introspection), are examples of the less-restricted (or broader) practice of introspection, on which the Gestaltists relied heavily.

I classify the Gestalt approach, and the general Wundtian experimental approach (distinct from the assumptions of analytical introspection), as forms of introspection in a broad sense—forms of psychological investigation that are mediated by observers' or subjects' responses to their own experience. This form of introspection has limits, as Wundt knew. He realized that many processes cannot be fully observed in consciousness. We should not expect introspection to directly reveal the nature of mind or the structure of psychological processes.[16] The development of cognitive and perceptual psychology has only confirmed this limitation. Nevertheless, the perceived qualities of objects, the temporal structure of experience, and the effects of experimental manipulations on experience can be investigated by techniques that rely on subjects' responses to what they experience, and they can be used as evidence in investigating the structure of underlying psychological processes.

INTROSPECTIVE REALITY

There are many examples of the use of introspection (broadly construed) in present-day psychology. Every textbook in perception employs demonstration drawings, sometimes similar to those used by the Gestaltists, to illustrate various perceptual phenomena. All of them depend upon the reader's being able to attend to the way the drawing looks, and to recognize appropriate aspects of how he or she experiences the drawings, including, in the case of figure/ground reversal, changes in phenomenal organization that occur while the physical object (the line drawing itself) remains the same on the printed page.

I want to examine some experiments in which subjects attend to or respond to aspects of their occurrent experience. I describe some psychological experiments on shape perception in some detail, and also mention some work on color perception and attention.

One phenomenon studied in perceptual psychology is *shape constancy*, the tendency of objects to appear to have a constant shape despite differences in viewing conditions (especially viewing angle, in the case of flat objects). Consider again a circular dinner plate. It appears circular when viewed at various angles, say, from 45° through 90°. At 90°, perpendicular to the line of sight, the plate projects a circular shape on the retina; at 45°, an ellipse (as at other angles, until the plate is seen edge-on, when it flattens to a long, thin shape with parallel edges, perhaps half-rounded at each end). In studies of shape constancy, the aim may be to distinguish the conditions in which full (or nearly full) constancy is obtained from those in which perception tends toward projective shape. One such set of conditions might include brief exposure to a set of stimuli that are generated as projectively equivalent shapes when viewed at predetermined angles (e.g., a circle viewed perpendicularly, and various ellipses that project a circle when rotated to various angles, say, 39°, 52°, and 65°). Experimenters may then elicit reports of perceived shape by, for example, asking subjects to pick out the one shape on a sheet of comparison shapes that most closely matches the perceived shape of an object they've just seen.

In studying shape constancy, experimenters have discovered that it is important to instruct observers concerning their attitude about what they are to report (Epstein, Bontrager, and Park, 1963). If observers believe that their job is to report the *perspective projection* of a shape, their reports will deviate from shape constancy (except at 90°). But such deviation may simply reflect their attitude about the task, not their perceptual experience of the shape. If subjects believe that they are to report what the actual or *objective shape* is, they may "correct" the appearance; under conditions of brief exposure, they may try to guess the objective shape. This could lead to reports closer to shape constancy than their perceptual experience would warrant. In consequence, subjects are typically instructed to report *phenomenal shape,* as opposed to *projective* or *objective* shape.

Subjects are instructed to base their report on what the shape of the object looks to be, not what they would guess it to be, nor what they think it should be.[17] Given such instructions, subjects have been found to report good shape constancy under conditions of binocular viewing (using both eyes, and without moving the head), when viewing an object illuminated for less than one-fifth of a second, followed by darkness. When their viewing is interrupted by a visual "mask" (small, irregular white shapes on black) at very brief periods (from 0 to 50 milliseconds) after offset of the illumination on the object, they tend to report projective shape (Epstein, Hatfield, and Muise, 1977). They also tend to report projective shape when viewing the shapes monocularly, that is, with one eye and no head movement (Epstein and Hatfield, 1978).

We need not enter into the theoretical significance of these reports. What is interesting to note is that, under instructions to report the shape as it appears, subjects exhibited shape constancy when the stimulus object was illuminated for less than one-fifth of a second, and they tended toward projective shape when uninterrupted viewing time was very brief, or when binocular depth information was eliminated. These findings are consistent with the conclusion that the observer's experience of the same objective shape at the same slant changed under differing conditions of observation. The changes are in the direction expected by theory. Hence, the consistency of the data suggests that these techniques, which draw on subjects' responses as mediated by their attention to their phenomenal experience, allowed experimenters to study aspects of that experience.

Color perception has been studied in the laboratory for more than 150 years. The methods of study, called "psychophysics," have been highly refined. Palmer (1999: 665) defines psychophysics as "the behavioral study of quantitative relations between people's perceptual experiences and corresponding physical properties." The studies are behavioral because they rely on subjects' responses, whether verbal (saying "yes" or "no") or manual (pressing a button, adjusting a dial). They depend on perceptual experience because they concern color appearances. In studies of color matching (Kaiser and Boynton, 1996: 124–25), subjects may look at a round area or disc that is illuminated by two different sources. On the left hemifield, a monochromatic light of known wavelength is projected. On the right, a mixture of two monochromatic lights is projected. The subject is asked to vary the mixture by turning a knob until the disc appears uniform (no border or difference between the two hemifields is apparent). The subjects' responses are mediated by the appearance of the disc: how it looks to them. The resulting color matches are among the fundamental data for color theory. The results are highly consistent (with very tight error bars) for normal human observers (normal trichromats).

Finally, work on attention has blossomed in the experimental literature in recent decades. Many techniques are used to measure the effects of

attention, either directly or indirectly. Indirect measures may include subjects' abilities to report one sort of thing while attending to something else
(say, to report on the shapes of objects they saw recently, when they had
only been told to look for a specific color).[18] One striking technique, which
has been used to study attentional processes, is called "pop out." It is based
on the phenomenological observation that in a field of uniform objects, say,
letters or shapes, even under brief exposure (too brief to allow eye movement or redirection of attention), a non-matching shape (calibrated to be
of similar size and color to surrounding shapes) will "pop out" or become
phenomenally salient. If subjects are asked to detect the presence or
absence of a diagonal line-segment among vertical line-segments of the
same length, the time it takes them to do so is not affected by the number
of vertical segments (ranging from 2 to 32). This finding led Triesman and
Gelade (1980) to conclude that subjects process the identity of such shapes
preattentively and in parallel (that is, all at once, as opposed to serially, one
after another). Details of the studies aside, the important point is that subjects' responses are mediated by directed attention and phenomenal
salience. Here again, experimental psychology uses introspectively mediated responses that exactly fit Wundt's broad conception of introspection.

279

INTROSPECTION AS EVIDENCE

When introspection is defined as deliberate and immediate attention to certain aspects of phenomenal experience, we see that it continues to be used
as a source of evidence in perceptual and cognitive psychology. The psychologists who use it need not be, and often are not, committed to the existence
of distinct entities that, like sense-data, have phenomenal properties of their
own, distinct from those involved in the direct perception of external
objects. The key to introspection is not "looking within," but attending to
relevant aspects of experience. Such relevant aspects include phenomenal
variations in the looks of things. These variations may be at the coarse grain
of object description ("the thing looks red"). For the purposes of perceptual
psychology, however, the concepts involved will classify how things look at a
finer grain of description than is used in ordinary typing of objects and their
properties. Introspectively based responses may require persons to attend
more closely to phenomenal shape than they normally do, or to inspect
shades of color more closely than they usually do (except, perhaps, in the
paint store).

Such responses are treated as scientific evidence in the literature of
experimental psychology. The evidence purports to reveal facts about
attention, or shape perception, or color perception in general. One person's introspectively based response is treated as yielding information

about how others will respond as well—subject to known, or discovered, individual differences (as in color blindness).

This literature shows little or no concern with an epistemological worry raised by philosophers: that introspection is inherently "private" or "subjective." When philosophers make this objection, they may contrast the alleged privacy of introspection with the epistemologically more worthy perception of, or response to "public" objects, such as tables and chairs. As philosophers often conceive these things, tables and chairs have properties that all of us can perceive and compare, by contrast with sense-perceptions themselves, which are private to each subject. I can't have yours, you can't have mine, so we can't check or compare them.

This framing of the problem of privacy retains the earlier confusion about what the object of introspection is supposed to be in perceptual psychology. In the standard case, the focus of attention is how the distal object looks. In fact, knowledge of the "public" object depends on the same phenomenal experience. There are not two experiences: one of the table as public, one of the experience of the table as private.[19] The only difference between objective property reports and introspective reports are the concepts that are used to classify the experience. In the first case, the subject has learned to attribute determinate properties of color and shape that are counted as remaining the same under large variations in lighting and in viewing distance. We know to expect that the table's shape and color are stable. In the second case, we are interested in subtler variations in phenomenal color and shape. We may be describing the same table, looking exactly the same, in the two cases. But the concepts are of different grain and application. We classify a table we've just painted as "a uniform red across its surface" when we apply object-color concepts. But we may describe variations in the appearance of the uniformly red pigment (due to lighting variation, shadows, glare, etc.) when we adopt an attitude of phenomenal description.

It is true that two observers can't directly compare their phenomenal experiences in such cases. But they can't directly compare how they perceive the table as an external object, either. In both cases, we as observers coordinate our descriptions with repeated samplings of how the table looks, and we develop language for conveying those looks, with the stable table as the coordinating factor. There need be no mystery in this, as Köhler (1947: 19–33) has explained. The physical world itself is known to us directly only by our experiencing it—visually, according to how it looks, and tactually, according to how it feels.[20]

Introspection may be taken as a reliable source of data about objects of consciousness. In perception, introspectively based responses go to how things look. These responses provide data about phenomenal experience, and such data are legitimate objects of explanation in perceptual psychology. Introspectively based responses are no longer considered to provide direct access to the structure and functioning of the psychological processes that underlie visual perception. Rather, these processes must be inferred from, or hypotheses about them must be tested against, various patterns of data. The relevance of introspection for discovering fundamental psychological processes has been reevaluated more than once during the past century. Introspection can yield data to mediate inferences or to test hypotheses. It is not and need not be seen as an oracle whose pronouncements can, by their immediacy, lay psychological processes bare. It provides evidence, and that's all. But that should be plenty.[21]

281

NOTES

1. Hume (1739–40: 252): "when I enter most intimately into what I call myself, I always stumble on some particular perception or other, of heat or cold, light or shade, love or hatred, pain or pleasure. I never can catch myself at any time without a perception, and never can observe any thing but the perception." Kant (1781/1787: A 355): "It is however obvious that through the I attached to thought the subject of inherence is designated only transcendentally, without noting any quality in it whatsoever, or in general being acquainted with or knowing anything from or of it"; also Kant (1781/1787:, A 346/B 404, A 360, B 408, 420). As is usual, "A" and "B" refer to the original pagination of the first and second editions of Kant (1781/1787).

2. Neither philosopher ascribes this view directly to an opponent, but they do make clear that it would be inappropriate to view thoughts or the soul as mathematical points or as directly perceivable objects. Hume (1739–40: 235): "Neither ought a desire, tho' indivisible, to be consider'd as a mathematical point." Kant (1783/2004: 90): "to think of the soul as a simple substance already amounts to thinking of it as an object (the simple) the likes of which cannot at all be represented to the senses." A mathematical point is not visible, hence could not be seen; Kant might instead be alluding to the idea that "things in themselves" have no spatial properties at all, hence cannot even be described as points.

3. In his correspondence, Descartes (1991: 306, 354–55, 356–57) distinguished between merely having a sensation, and reflecting upon, or becoming aware of, the facts about and characteristics of the sensation (or other act of mind). Both sorts of mental state are "conscious," but the first sort may not happen to be remembered. Further, the reflective act is performed by the "pure intellect," and it is characterized as a "perception" that takes the sensation (or other act of mind) as its object. These reflective acts are akin to "introspection" considered as the perception of facts about a mental state. I discuss Descartes on consciousness in Hatfield (2003a: 122–25, 325–27).

4. I leave aside the question of whether Wittgenstein (1953) actually intended his "private language argument" to tell against introspective awareness; the matter is under dispute (Sluga, 1996).

5. This introduces the notion of a phenomenal-report sense of "looks," which is distinct from Austin's notion of "looks" as introducing doubt about the reality of things or the reliability of current perception. Mundle (1971: 15–20) admirably defends a phenomenological sense of "looks." I must confess, however, that I disagree with some of his phenomenal reports, as when he takes "perspectival shape" to indicate the "real" looks of things (1971: 27–28).

6. In the technical literature of color perception, some theorists use the term "lightness" for the perceived object color of the paper itself (as white or gray), and "brightness" for the phenomenal variations of darker or lighter white (Rock, 1975: 503–04). In this technical context, it is incorrect to say the paper looks "grayer" in a certain region, when this is not intended to ascribe an object color (pigmentation) to that region. But in nontechnical language, shadow may be described as "graying" a region of the paper.

7. Our authors offer various views on introspection, but Rey (1997: 136–7) presents the "transparency" point as a problem for the phenomenology of introspection. Harmon (1990) is directly negative about the possibility of introspection. Dretske (1995, ch. 2) allows introspection in the form of beliefs about the content of perception, but thinks it can have no experiential content or phenomenology peculiar to it. Tye (1995; 2002) is the most liberal in his willingness to countenance introspective experiences, but he goes to heroic lengths to preserve the theory that their content is exhausted by external-world content; thus, in the case of things that look blurry (due to nearsightedness, let us say), he maintains that the content is as of a vague or blurred object in the world, thereby avoiding ascribing any specifically subjective content (Tye, 2002). As stated in the text, my arguments in this paper supporting introspection do not rely on any particular conception of the metaphysics of sensory qualities; elsewhere, I support the view that subject-dependent phenomenal qualities exist (Hatfield, 2003c; 2004).

8. Factors such as sensory adaptation and habituation limit the extent to which "the same" phenomena can be observed over time and in repeated trials. Such effects can be controlled for, and may be counted as experimental error, or they can be studied in their own right (Palmer, 1999: 674).

9. Dretske puts the point as follows: "If there is an inner sense, some quasi-perceptual faculty that enables one to know what experiences are like by 'scanning' them, this internal scanner, unlike the other senses, has a completely transparent phenomenology. It does not 'present' experiences of external objects in any guise other than the way the experiences present external objects. If one is aware of experiences in the way one is aware of external objects, the experiences look, for all the world, like external objects. This is very suspicious. It suggests that there is not really another sense in operation at all" (1995: 62). However, the notion of an "inner sense," or of a "perceptual model" of introspection is ambiguous. Dretske and others have interpreted the analogy between introspection and sensing or perceiving to mean that there must be a distinct object of sense, and that "perceiving" such objects introspectively must presuppose a perceiver-perceived relation distinct from that already extant in the perception of external objects. But one may interpret the notion of "perception" more broadly. If we include cognitive aspects of perception (Palmer, 1999: 13), such as classifying objects (seeing something *as a book*, rather than simply seeing its shape and color), as part of the perceptual act (and its phenomenology), then the "perceptual model" of introspection can be interpreted as the application of introspective concepts within everyday perceptual experience. To introspect would be to apply concepts in classifying one's immediate experience, for example, to conclude that the paper looks grayish here and reddish there, where this classifying is understood to be distinct from ascribing a surface-color property (a pigmentation) to the paper itself.

10. In saying that a sensation is not a "different object" from external objects, Wundt may appear to be taking a metaphysical stance and adopting a form of monism. In fact, he preferred not to adopt a metaphysical "hypothesis" on the mind-body problem; in equating "inner" and "outer" sense perceptions as objects of experience, he was not making an assertion about their ontological status. Wundt himself subscribed to a form of

psychological parallelism as a methodological principle, but he did not purport to refute materialism or idealism (spiritualism); rather, he characterized them as empirically sterile (1901/1902: 352–63). In this regard, he shared the position of Helmholtz (and others in the late nineteenth century) that the data of science are the materials of observation, and that these can be known independently of "metaphysical" notions such as material or mental substance. One might suspect that this position leads to the "neutral monism" common to Mach, James, and Russell, which itself may be thought to tend toward phenomenalism or idealism; but the view that perceptual experience can be investigated independently of a particular position on the mind-body problem can be defended without subscribing to neutral monism (see Hatfield, 2004, and sec. 4 below).

11. One can of course also study afterimages and other subjective phenomena, but let us stay focused on the primary case.

12. Some objects of introspective observation may be spatially "inner" in the sense of inside the skin, as a pain in the stomach is. Others, such as anger or joy, may be ascribed as feelings of the person, generally felt as localized in the region of the body. One wouldn't try to turn one's eyeballs around to see them; one would simply reflect on the character of one's emotions and feelings.

13. Early on, Wundt (1862) adopted a "punctiform" analysis of perception (see Hatfield, 1990, ch. 5), and he retained sympathy for a "fusional" account of the origin of spatial structure from elemental or atomic sensations (1901/1902: 116–56), and of the development of tones from constituent sensations (100–113). Titchener (1910: 304–05) did not follow Wundt on the original non-spatiality of visual sensations.

14. On "analytical introspection" versus "phenomenological" or "naive" introspection, see Rock (1975: 11–12). Koffka (1935: 73) distinguished phenomenological introspection from the "American" version of introspection (i.e., analytical introspection). Palmer (1999: 48) attributes an "introspective" approach to the Gestaltists, owing to their appeal to "phenomenological observations of one's own conscious experience." He distinguishes this sort of introspection from that involved in the search for sensory atoms, which he (somewhat unhappily) calls "trained introspection" (1999: 50). Many psychologists still associate the term "introspection" with analytical introspection. I term this the "narrow" conception, and distinguish it from a broader notion (citing, if needed, the precedents just given from past and present literature).

15. Here I adopt the usual description of shape constancy as yielding a Euclidean circle (for a round dinner plate). In fact, the space of visual constancy may be compressed with distance, so that the plate would be represented with a slight flattening of the front and rear edges—which would nonetheless be taken for the look that a true circle should have. On the compression of visual space, see Hatfield (2003c).

16. The Gestaltists thought it might, through an isomorphism relation, but in the form they held it this notion has been discredited (see Epstein and Hatfield, 1994).

17. In Epstein, Hatfield, and Muise (1977), subjects were instructed as to the purpose of the experiment, and about the kind of reports they were to make (how they should conceive the task). They were first told: "In this experiment we are trying to learn how the apparent shape, or the appearance of the shape, of an object is affected by variations in the time one is allowed to see the object." After a description of the experimental setup, and prior to explaining how responses were to be indicated, they were told: "I would like to make clear what it is that I am asking you to report. I want you to report the shape of the object directly as it appears to you, without any analysis or guessing on your part. The experiment will be spoiled if you base your response on a conscious attempt to figure out what the shape *ought* to be, instead of what it appears to be. Don't convert the situation into

a guessing game or into an intellectual task. We are not trying to trick you in this experiment; we are really interested in the way things look to you." Similar instructions were given for trials on which slant (rather than shape) judgments were elicited.

18. In the terminology of this chapter, an experiment in which subjects are primed to look for color but then queried on shape is an example of both introspection (the stimulus attribute they are directed to attend to) and inner perception (a dimension they are subsequently asked to report on). The first fits the notion that psychological introspection is a form of observation. The second presumably relies on memory of the undirected or "incidental" awareness of shape.

19. This does not deny, of course, that differences in attitude *can* (in some cases) cause the experience itself to differ, as when one takes what was described in the previous section as a "projectivist" attitude, or when one causes a figure/ground shift by redirecting attention. The point is that an introspective attitude need not change the spatial or chromatic character of experience. Nor need it be directed at a different experience of spatial and chromatic properties than that which occurs when one is observing an object without an introspective attitude. Moreover, introspecting may change the overall experience (by injecting a different conceptualization, one based on phenomenal concepts), without necessarily affecting spatial and chromatic characteristics.

20. I discuss these epistemological worries more fully in Hatfield (2004).

21. I am grateful to Yumiko Inukai, Jeffrey Scarborough, and Morgan Wallhagen for comments on an earlier version of this chapter.

REFERENCES

Armstrong, D. M. (1980) "What Is Consciousness," in his *Nature of Mind and Other Essays*, 55–67. St. Lucia: University of Queensland Press.

Brentano, Franz Clemens (1874) *Psychologie vom empirischen Standpunkte*. Leipzig: Duncker & Humblot. Trans. by Antos C. Rancurello, D. B. Terrell, and Linda L. McAlister, *Psychology from an Empirical Standpoint*. London: Routledge, 1995.

Broad, C. D. (1923) *Scientific Thought*. London: Kegan Paul, Trench, Trubner & Co.

Comte, Auguste (1830–42) *Cours de philosophie positive*. Paris: Rouen. Trans. by Harriet Martineau, ed., *The Positive Philosophy*. New York: Calvin Blanchard, 1855.

Descartes, Rene (1641) *Meditationes de prima philosophiae*. Paris: Soly. Trans. as "Meditations on First Philosophy," in John Cottingham, Robert Stoothoff, and Dugald Murdoch, eds., *Philosophical Writings of Descartes*, Vol. 2. Cambridge: Cambridge University Press, 1984.

Descartes, Rene (1644) *Principia philosophiae*. Amsterdam: Elzevir. Trans. as "Principles of Philosophy," in John Cottingham, Robert Stoothoff, and Dugald Murdoch, eds., *Philosophical Writings of Descartes*, Vol. 1. Cambridge: Cambridge University Press, 1985.

Descartes, Rene (1991) *Correspondence*, in John Cottingham, Robert Stoothoff, Dugald Murdoch, and Anthony Kenny, eds., *Philosophical Writings of Descartes*, Vol. 3. Cambridge: Cambridge University Press.

Dretske, Fred (1995) *Naturalizing the Mind*. Cambridge, Mass.: MIT Press.

Epstein, William, H. Bontrager, and J. Park (1963) "The Induction of Nonveridical Slant and the Perception of Shape," *Journal of Experimental Psychology* 63: 472–79.

Epstein, William, and Gary Hatfield (1978) "Functional Equivalence of Masking and Cue Reduction in Perception of Shape at a Slant," *Perception and Psychophysics*, 23: 137–44.

Epstein, William, and Gary Hatfield (1994) "Gestalt Psychology and the Philosophy of Mind," *Philosophical Psychology*, 7: 163–81.

Epstein, William, Gary Hatfield, and Gerard Muise (1977) "Perceived Shape at a Slant as a Function of Processing Time and Processing Load," *Journal of Experimental Psychology: Human Perception and Performance*, 3: 473–83.

Harman, Gilbert (1990) "The Intrinsic Quality of Experience," in J. E. Tomberlin, ed., *Philosophical Perspectives*, Vol. 4, "Action Theory and Philosophy of Mind," pp. 31–52. Atascadero, Calif.: Ridgeview.

Hatfield, Gary (1990) *The Natural and the Normative: Theories of Spatial Perception from Kant to Helmholtz*. Cambridge, Mass.: MIT Press.

Hatfield, Gary (2003a) *Descartes and the Meditations*. London: Routledge.

Hatfield, Gary (2003b) "Psychology Old and New," in Thomas Baldwin, ed., *Cambridge History of Philosophy, 1870–1945*, pp. 93–106. Cambridge: Cambridge University Press, 2003.

Hatfield, Gary (2003c) "Representation and Constraints: The Inverse Problem and the Structure of Visual Space," *Acta Psychologica*, 114: 355–78.

Hatfield, Gary (2004) "Sense-Data and the Mind-Body Problem," in Ralph Schumacher, ed., *Perception and Reality: From Descartes to the Present*, pp. 305–31. Berlin: Mentis Verlag.

Hume, David (1739–40) *Treatise of Human Nature*. London: John Noon.

Hume, David (1748) *Philosophical Essays Concerning Human Understanding*. London: Millar. As reprinted in Tom L. Beauchamp, *An Enquiry Concerning Human Understanding*. Oxford: Oxford University Press, 1999.

Jackson, Frank (1998) *From Metaphysics to Ethics: A Defence of Conceptual Analysis*. Oxford: Clarendon Press.

James, William (1890) *Principles of Psychology*, 2 vols. New York: Holt and Co.

James, William (1912) "The Experience of Activity," in his *Essay in Radical Empiricism*, pp. 155–189. New York: Longmans, Green, and Co.

Kaiser, P. K., and R. M. Boynton (1996) *Human Color Vision*, 2nd edn. Washington, D.C.: Optical Society of America.

Kant, Immanuel (1781/1787) *Kritik der reinen Vernunft*, 1st and 2nd edns. Riga: Hartnoch. (Translations in the text are by the author.)

Kant, Immanuel (1783) *Prolegomena zu einer jeden künftigen Metaphysik*. Riga: Hartnoch. Trans. by Gary Hatfield, ed., *Prolegomena to Any Future Metaphysics*. Cambridge: Cambridge University Press, 2004.

Koffka, Kurt (1935) *Principles of Gestalt Psychology*. New York: Harcourt, Brace.

Köhler, Wolfgang (1947) *Gestalt Psychology*. New York: Liveright. (An earlier English edition was published in 1929, and a German edition in 1933.)

Kusch, M. (1999) *Psychological Knowledge: A Social History and Philosophy*. London: Routledge.

Lycan, William (1997) "Consciousness as Internal Monitoring," in Ned Block, Owen Flanagan, and Guven Guzeldere, eds., *Nature of Consciousness*, pp. 755–71. Cambridge, Mass.: MIT Press.

Lyons, William (1986) *Disappearance of Introspection*. Cambridge, Mass.: MIT Press.

Mill, John Stuart (1865) *Auguste Comte and Positivism*. London: Trübner.

Moran, Richard (2001) *Authority and Estrangement: An Essay on Self-Knowledge*. Princeton: Princeton University Press.

Mundle, C. W. K. (1971) *Perception: Facts and Theories*. Oxford: Oxford University Press.

285

Myers, Gerald E. (1986) "Introspection and Self-Knowledge," *American Philosophical Quarterly*, 23: 199–207.

Palmer, Stephen E. (1999) *Vision Science: Photons to Phenomenology*. Cambridge, Mass.: MIT Press.

Rey, Georges (1997) *Contemporary Philosophy of Mind: A Contentiously Classical Approach.* Cambridge, Mass.: Blackwell.

Rock, Irvin (1975) *Introduction to Perception.* New York: Macmillan Publishing Co.

Russell, B. (1919) "On Propositions: What They Are and How They Mean," *Proceedings of the Aristotelian Society*, suppl. vol. 2, *Problems of Science and Philosophy*, pp. 1–43. Republished in Russell, *Logic and Knowledge: Essays, 1901–1950*, pp. 285–320. New York: Macmillan, 1956.

Shoemaker, Sydney (1996) *First-Person Perspective and Other Essays.* Cambridge: Cambridge University Press.

Skinner, B. F. (1963) "Behaviorism at Fifty," *Science*, 140: 951–58.

Sluga, Hans (1996) "'Whose House Is That?': Wittgenstein on the Self," in Hans Sluga and David G. Stern, eds., *Cambridge Companion to Wittgenstein*, pp. 320–53. Cambridge: Cambridge University Press.

Titchener, E. B. (1910) *A Text-Book of Psychology.* New York: Macmillan.

Titchener, E. B. (1912a) "Prolegomena to a Study of Introspection," *American Journal of Psychology*, 23: 427–48.

Titchener, E. B. (1912b), "Schema of Introspection," *American Journal of Psychology*, 23: 485–508.

Triesman, Anne, and G. Gelade (1980) "A Feature Integration Theory of Attention," *Cognitive Psychology*, 12: 97–136.

Tye, Michael (1995) *Ten Problems of Consciousness: A Representational Theory of the Phenomenal Mind.* Cambridge, Mass.: MIT Press.

Tye, Michael (2002) "Representationalism and the Transparency of Experience," *Noûs* 36: 137–51.

Watson, J. B. (1914). *Behavior: An Introduction to Comparative Psychology*, New York: Holt.

Wittgenstein, Ludwig (1953). *Philosophische Untersuchungen.* New York, Macmillan. Trans. by G. E. M. Anscombe, ed., *Philosophical Investigations* (on facing pages).

Wundt, Wilhelm (1882) "Die Aufgaben der experimentellen Psychologie," *Unsere Zeit*; as reprinted in his *Essays*, pp. 127–53. Leipzig: Engelmann, 1885.

Wundt, Wilhelm (1888) "Selbstbeobachtung und innere Wahrnehmung," *Philosophische Studien* 4: 292–309.

Wundt, Wilhelm (1901) *Grundriss der Psychologie*, 4th edn. Leipzig: Engelmann. Trans. by C. H. Judd, ed., *Outlines of Psychology.* Leipzig: Engelmann, 1902.